Methods in Molecular Biology

Series Editor
John M. Walker
School of Life and Medical Sciences
University of Hertfordshire
Hatfield, Hertfordshire, UK

For further volumes:
http://www.springer.com/series/7651

For over 35 years, biological scientists have come to rely on the research protocols and methodologies in the critically acclaimed *Methods in Molecular Biology* series. The series was the first to introduce the step-by-step protocols approach that has become the standard in all biomedical protocol publishing. Each protocol is provided in readily-reproducible step-by-step fashion, opening with an introductory overview, a list of the materials and reagents needed to complete the experiment, and followed by a detailed procedure that is supported with a helpful notes section offering tips and tricks of the trade as well as troubleshooting advice. These hallmark features were introduced by series editor Dr. John Walker and constitute the key ingredient in each and every volume of the *Methods in Molecular Biology* series. Tested and trusted, comprehensive and reliable, all protocols from the series are indexed in PubMed.

Inositol Phosphates

Methods and Protocols

Edited by

Gregory J. Miller

Department of Chemistry, The Catholic University of America, Washington, DC, USA

Humana Press

Editor
Gregory J. Miller
Department of Chemistry
The Catholic University of America
Washington, DC, USA

ISSN 1064-3745 ISSN 1940-6029 (electronic)
Methods in Molecular Biology
ISBN 978-1-0716-0166-2 ISBN 978-1-0716-0167-9 (eBook)
https://doi.org/10.1007/978-1-0716-0167-9

This Humana imprint is published by the registered company Springer Science+Business Media, LLC, part of Springer Nature.
The registered company address is: 233 Spring Street, New York, NY 10013, U.S.A.

Preface

In the 1980s, inositol 1,4,5 trisphosphate was discovered to participate in the mobilization of intracellular calcium and has since become an archetype of second messenger signaling molecules. Since the description of its first cellular roles, inositol phosphate signaling has expanded in every direction. The number of differently phosphorylated inositol molecules that have been identified to exist has increased greatly and ranges from one to as many as thirteen phosphates around the small six-carbon ring. This large network of molecules has been discovered to play a broad range of roles in cells of many species. These functions range from controlling enzyme and membrane channel activities, contributing to the proper folding of proteins, aiding in the accurate formation and targeting of intracellular vesicles, among many others. In parallel with the discoveries of these many functional roles, technical advances and chemical tools have been developed that aid in the identification and synthesis of variously phosphorylated inositol molecules and the impact of this family of molecules on their cellular environments.

This volume begins with a perspective on the history of methodologies to study inositol phosphates written by Stephen Shears, which sets our stage for a collection of methods that cover broad areas of this field of research. Each subsequent chapter reports a method and begins with a description of the use for the protocol and offers lists of all the reagents, materials, and equipment required to perform each of the protocols. The methods in this volume are divided into broad categories, beginning with isolation, detection, and measurement of inositol phosphates, followed by methods to determine the functions of inositol phosphates and the enzymes that generate them, and finally, chemical tools using inositol phosphates. The methods presented are diverse as are the functions of these fascinating signaling molecules, and we hope this collection of protocols may be useful in the further expansion of this field.

Washington, DC, USA *Gregory J. Miller*

Contents

Contributors

RASHNA BHANDARI • *Laboratory of Cell Signalling, Centre for DNA Fingerprinting and Diagnostics (CDFD), Hyderabad, India*

MARCO BIANCUCCI • *Department of Microbiology-Immunology, Feinberg School of Medicine, Northwestern University, Chicago, IL, USA; GSK Vaccines, Rockville, MD, USA*

MANASA CHANDURI • *Laboratory of Cell Signalling, Centre for DNA Fingerprinting and Diagnostics (CDFD), Hyderabad, India*

BRADLEY P. CLARKE • *Department of Biochemistry, Vanderbilt University School of Medicine, Nashville, TN, USA*

YANN DESFOUGÈRES • *Medical Research Council Laboratory for Molecular Cell Biology, University College London, London, UK*

QUYNH H. DUONG • *Department of Food Science and Technology, College of Agricultural and Environmental Sciences, The University of Georgia, Athens, GA, USA*

ANDREW T. HALE • *Department of Biochemistry, Vanderbilt University School of Medicine, Nashville, TN, USA*

XIN HUI • *Medical Faculty, Centre for Molecular Signalling (PZMS), Institute for Molecular Cell Biology and Research Center for Molecular Imaging and Screening, Saarland University, Homburg/Saar, Germany*

RAJAGOPAL SHARADA KILARI • *Department of Pharmacy, University of Wolverhampton, Wolverhampton, UK*

SUNG HYUN KIM • *Department of Physiology, School of Medicine, Kyung Hee University, Seoul, South Korea*

KEEHWAN KWON • *Infectious Disease Group, J. Craig Venter Institute, Rockville, MD, USA; Center for Structural Genomics of Infectious Diseases, Northwestern University, Chicago, IL, USA; GSK, Collegeville, PA, USA*

PETER LIPP • *Medical Faculty, Centre for Molecular Signalling (PZMS), Institute for Molecular Cell Biology and Research Center for Molecular Imaging and Screening, Saarland University, Homburg/Saar, Germany*

YASUO MORI • *Institute for Integrated Cell-Material Sciences, Kyoto University, Nishikyo-ku, Kyoto, Japan; Department of Synthetic Chemistry and Biological Chemistry, Graduate School of Engineering, Kyoto University, Nishikyo-ku, Kyoto, Japan*

TAKASHI MORII • *Institute of Advanced Energy, Kyoto University, Uji, Kyoto, Japan*

CANFANG NIU • *Key Laboratory for Feed Biotechnology of the Ministry of Agriculture, Feed Research Institute, Chinese Academy of Agricultural Sciences, Beijing, China*

RONALD B. PEGG • *Department of Food Science and Technology, College of Agricultural and Environmental Sciences, The University of Georgia, Athens, GA, USA*

ANA M. ROSSI • *Department of Pharmacology, University of Cambridge, Cambridge, UK*

STEPHEN T. SAFRANY • *RCSI-Bahrain, Adliya, Bahrain*

ADOLFO SAIARDI • *Medical Research Council Laboratory for Molecular Cell Biology, University College London, London, UK*

REIKO SAKAGUCHI • *Institute for Integrated Cell-Material Sciences, Kyoto University, Nishikyo-ku, Kyoto, Japan*

KARLA J. F. SATCHELL • *Center for Structural Genomics of Infectious Diseases, Northwestern University, Chicago, IL, USA; Department of Microbiology-Immunology, Feinberg School of Medicine, Northwestern University, Chicago, IL, USA*

STEPHEN B. SHEARS • *Laboratory of Signal Transduction, National Institute of Environmental Health Sciences, National Institutes of Health, Research Triangle Park, NC, USA*

SATOSHI SHUTO • *Laboratory of Organic Chemistry for Drug Development, Faculty of Pharmaceutical Sciences, Hokkaido University, Hokkaido, Japan*

MI SOOK SUNG • *National Creative Research Center for Single-Molecule Systems Biology, School of Biological Sciences, Seoul National University, Seoul, South Korea; Center for Theragnosis, Korea Institute of Science and Technology (KIST), Seoul, South Korea*

SHUNSUKE TAJIMA • *Institute of Advanced Energy, Kyoto University, Uji, Kyoto, Japan*

AKIHIKO TANIMURA • *Division of Pharmacology, School of Dentistry, Health Sciences University of Hokkaido, Hokkaido, Japan*

COLIN W. TAYLOR • *Department of Pharmacology, University of Cambridge, Cambridge, UK*

BENJAMIN L. TURNER • *Smithsonian Tropical Research Institute, Balboa, Ancon, Republic of Panama*

LUCINDA WINWARD • *Division of Cell Signalling, School of Life Sciences, University of Dundee, Dundee, UK*

PATRICK J. WOIDA • *Department of Microbiology-Immunology, Feinberg School of Medicine, Northwestern University, Chicago, IL, USA*

PEILONG YANG • *Key Laboratory for Feed Biotechnology of the Ministry of Agriculture, Feed Research Institute, Chinese Academy of Agricultural Sciences, Beijing, China*

BIN YAO • *Key Laboratory for Feed Biotechnology of the Ministry of Agriculture, Feed Research Institute, Chinese Academy of Agricultural Sciences, Beijing, China*

TAE-YOUNG YOON • *National Creative Research Center for Single-Molecule Systems Biology, School of Biological Sciences, Seoul National University, Seoul, South Korea; Center for Theragnosis, Korea Institute of Science and Technology (KIST), Seoul, South Korea*

JOHN D. YORK • *Department of Biochemistry, Vanderbilt University School of Medicine, Nashville, TN, USA*

Chapter 1

A Short Historical Perspective of Methods in Inositol Phosphate Research

Stephen B. Shears

Abstract

The multitudinous inositol phosphate family elicits a wide range of molecular effects that regulate countless biological responses. In this review, I provide a methodological viewpoint of the manner in which key advances in the field of inositol phosphate research were made. I also note some of the considerable challenges that still lie ahead.

Key words Inositol, Inositol phosphate, Inositol lipid, Phosphoinositide, Signaling, Methods, History, Calcium, Phospholipase C, HPLC, Kinases, Inhibitors, Radiolabeling, Mass-assays

1 Introduction

It is now 14 years since I first read a 2005 perspective by Michael Berridge [1] in which he illuminated the technical hurdles that were surmounted in order to demonstrate Ca^{2+} mobilization by Ins(1,4,5)P_3. He gave that story some perspective, by encouraging the contemporary reader not to view this discovery through the eyes of a researcher for whom "the experimental techniques to study the action of intracellular messengers are now relatively easy." I balked at that statement, since in 2005 it did not reflect my view of the ease of performing research in the field of inositol phosphates. I reacquainted myself with Mike's autobiographical account before beginning this article, and my opinion from 2019 has not softened.

I do not dispute that the newer technology underpinning today's signaling research has generally made lab-work less arduous and improved productivity. I am frequently reminded of that fact during our lab meetings, when discussion on the topic of introducing a new assay will inevitably generate the question: "Is there a kit for that?" That query is rarely appropriate for inositol phosphate research.

Gregory J. Miller (ed.), *Inositol Phosphates: Methods and Protocols*, Methods in Molecular Biology, vol. 2091,
https://doi.org/10.1007/978-1-0716-0167-9_1,

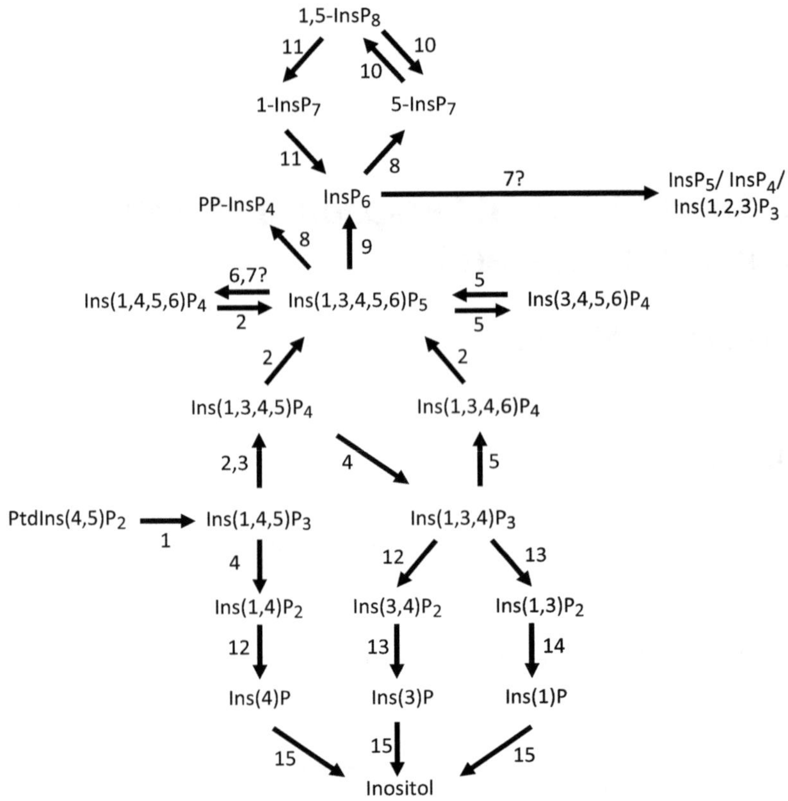

Fig. 1 Predominant pathways of inositol phosphate metabolism in mammals. Only the major mammalian pathways are shown; in other groups of organisms some of these enzymes have different catalytic preferences, or alternately are not present. Background information can be obtained from the following reviews [2–5]. Enzymes are numbered as follows: (1) Phospholipase C (PLC); (2) Inositol polyphosphate multikinase (IPMK); (3) Ins(1,4,5)P_3 3-kinase (IP3K); (4) Ins(1,4,5) P_3/Ins(1,3,4,5)P_4 5-phosphatase; (5) Inositol tetrakisphosphate 1-kinase (ITPK1); (6) Phosphatidylinositol 3,4,5-trisphosphate 3-phosphatase and dual-specificity protein phosphatase (PTEN); (7) multiple inositol polyphosphate phosphatase (MINPP1)—the question mark is to draw attention to the conundrum that this enzyme is compartmentalized inside endoplasmic reticulum, with no demonstrated access to its substrates; (8) Inositol hexakisphosphate kinase (IP6K); (9) Inositol 1,3,4,5,6-pentakisphosphate kinase (IP5K); (10) Diphosphoinositol pentakisphosphate 1-kinase (PPIP5K); (11) Diphosphoinositol polyphosphate phosphohydrolase (DIPP); (12) Ins(1,3,4)P_3/Ins(1,4)P_2 1-phosphatase; (13) Ins (1,3,4)P_3/Ins(3,4)P_2 4-phosphatase; (14) Ins(1,3)P_2 1-phosphatase; (15) Inositol monophosphatase (InsP phosphatase)

Even the mere quantification of cellular inositol phosphates continues not to be routine. For a start, there are an awfully large number of them to quantify (Fig. 1). Moreover, these molecules do not adsorb or emit UV or visible wavelength light, and so they are virtually undetectable by spectrophotometric or fluorescence methods. Instead, when assaying inositol phosphate turnover in intact cells, most in the field rely on [^{3}H]inositol radiolabeling, followed by HPLC-based separation of the [^{3}H]-labeled peaks of interest; the HPLC equipment is expensive, it requires careful maintenance to maintain optimal chromatography, and the methodology is decidedly low throughput. This is not a "relatively easy" technique.

There are other practical difficulties due to the inherent chemical nature of *myo*-inositol; the carbon ring has a plane of symmetry across its 2-5 axis, giving rise to an analytical complication that enantiomers of inositol phosphates are produced inside cells. To illustrate why that can be a problem, let us consider the enantiomeric pair, Ins(3,4,5,6)P_4 and Ins(1,4,5,6)P_4. These two molecules cannot be resolved from each other using standard, nonchiral chromatographic procedures. This complicates studies into the regulation of plasma membrane chloride fluxes by receptor-dependent changes in cellular Ins(3,4,5,6)P_4 levels [6]; the quantification of cellular Ins(3,4,5,6)P_4 is complicated by its "contamination" with Ins(1,4,5,6)P_4. Conversely, those investigating transcriptional regulation through changes in cellular levels of Ins (1,4,5,6)P_4 [7] will find *their* molecule of interest to be "contaminated" with Ins(3,4,5,6)P_4.

The plane of symmetry across the inositol ring also adds an extra challenge to their chemical synthesis. Such procedures generally require control over functional group positioning that must not only be regioselective but also enantioselective. It is also difficult to assign the absolute configuration of the synthetic inositol phosphate exclusively by measurement and comparison of optical rotations, which are small in value, and highly sensitive to contaminants and pH [8]. Another problem with *myo*-inositol being a meso compound is that functional experiments with inositol phosphates will sometimes be confounded by their interacting with the binding sites of receptors and enzymes in more than one orientation (i.e., "mode"). This enables one InsP to imitate another by presenting to the docking site the same, key recognition features [9]. For example, draw molecules of Ins(1,3,4)P_3 and Ins(3,4,5,6)P_4 so that the axial 2-OH group of both is positioned to the right [2]. They look rather different, until the Ins(1,3,4)P_3 is flipped vertically and rotated, whereupon its substituents at carbons 1, 3, 4, 6 spatially substitute for those of carbons 6, 4, 3, and 1 in Ins(3,4,5,6)P_4 [2]. Before my own group fully appreciated the consequences of this mimicry, we vainly attempted to purify and sequence the Ins (3,4,5,6)P_4 1-kinase [10], only to subsequently realize [11] it was the very protein that had previously been cloned as an Ins(1,3,4)P_3 5/6-kinase [12].

I will return later to some of technical challenges mentioned above, as I elaborate on how, over many years, methodological advances have facilitated the analysis of the structure, metabolism and function of inositol phosphates. I hope that approach also serves to underscore the value of the chapters provided by other contributors to this book. But please bear in mind this is not supposed to be an exhaustive history of inositol phosphate research, but instead, a methodological perspective on how selected, key advances were achieved.

2 The Early Years, Beginning with $InsP_6$

Anyone looking back over the methodology applied to the first 80 to 100 years of inositol phosphate research will appreciate the glacial nature of its pace. Once a particular method was introduced, it would generally exhibit remarkable longevity. A classic example is "Scherer's test" from 1850 [13]: the characteristic brick-red color of the calcium salt of rhodizonic acid that is formed upon oxidation of inositol with concentrated nitric acid. Despite its unsophisticated and relatively insensitive nature, it remained in use well into the 1950s [14].

It was Scherer's test that was used in the early 1900s to determine that inositol is a constituent of a phosphorylated molecule that accounts for 80% of the organic phosphate in plant seeds [15–18]. The magnesium and calcium salt of this new organic polyphosphate in plants became known as *phytin*, the Anglicized form of *la phytine*, the term originally coined by Swigel Posternak in 1903 [16]. Eventually, phytic acid became the more popular moniker for this molecule.

Posternak [16] was the first to accurately determine this new molecule comprised a 1:1 phosphate–carbon ratio, although he initially concluded that *la phytine* was a two-carbon diphosphate. The concept that the compound is actually a hexakisphosphate ($InsP_6$) was subsequently assisted by the demonstration that its enzymatic dephosphorylation yields the six-carbon inositol scaffold [17, 18]. Thus, the enzymatic activity in plants that degraded it was introduced as "phytase" [18]. We can therefore point to that particular discovery as representing the birth of scientific interest in inositol phosphate *metabolism*. As for early thoughts on the actual role of $InsP_6$ in plant seeds, it seemed obvious it must be a means to store phosphate until germination.

These determinations that plant seeds contain high amounts of $InsP_6$ naturally generated interest in its nutritional significance. To pursue this topic, a reasonable assay was required. In a 1914 paper, $InsP_6$ was acid-extracted from cereal, and assayed by using ammonium thiocyanate as an internal colorimetric indicator of the end-point of titration of the $InsP_6$ with Fe^{3+} [19]. This method for quantification of $InsP_6$, with only minor modifications, remained popular into the 1930s. By that time it was established that dietary $InsP_6$ is poorly digested by humans and other monogastric mammals; much of what is ingested is excreted in the feces [20]. The conclusions to derive from the use of such simple methodology remain momentous to this day: "factory farming" of pigs and poultry requires huge quantities of animal feed—soybeans, grains, and other seeds—but much of the $InsP_6$ in this food is excreted. In order to ensure that adequate dietary phosphate *is* available, it is added to the feed in its inorganic form, thereby accelerating consumption of the planet's nonrenewable phosphate reserves [21]. Excess, excreted inorganic phosphate, plus the $InsP_6$, enters the run-off from these farms and imperils adjacent waterways with eutrophication [21]. Hence, the considerable agricultural interest in bioengineering phytase to increase its activity and stability, so that it may be used in animal feed to release phosphate from $InsP_6$. This is a topic that is covered by Niu et al., in Chapter 14.

The first reports that $InsP_6$ might also be present in an animal cell were published around 1940. Rapoport's group [22, 23] reported that nucleated erythrocytes of birds and turtles contain substantial quantities of this polyphosphate. For these experiments, they used Scherer's inositol test again, plus a 1925 method to assay phosphate [24]. Unfortunately, these observations did not attract much attention, and so it took another 30 years before follow-up work found that the $InsP_6$ was a misidentification; the original determination of the phosphate:inositol ratio had been incorrect [25]. The true nature of that polyphosphate in erythrocytes was then determined to be $Ins(1,3,4,5,6)P_5$ [25].

$InsP_6$ continued to interest plant biochemists; the idea that $InsP_6$ is a phosphate-storage molecule led to studies into its synthesis and metabolism. These goals required the separation of inositol phosphates based on the number of phosphate groups that they contained. By the end of the 1950s, three main techniques had been developed. Two of these are paper chromatography [26, 27], and paper electrophoresis [28]; inositol phosphates were detected post-chromatography with an acid-molybdate reagent [29] or by fixation with ferric ion followed by treatment with salicylsulfonic acid [30]; both procedures have a sensitivity of 30-60 nmol Pi. For those wishing to also measure the inositol content of their preparations, Scherer's test had by that time finally been supplanted by a new method that relied upon recording the amount of periodate required to oxidize a sample of inositol. However, mg quantities of starting material were required [31–33].

Gravity-fed ion-exchange chromatography was also introduced to separate and resolve inositol phosphates [34–36]; fractions containing different inositol phosphates were concentrated and then boiled in acid to liberate free Pi for assay. Ion-exchange chromatography has persevered to this day as a convenient method to assay PLC activity, although most now quantify radiolabeled inositol phosphates, typically after their extraction from [^{3}H]inositol-labeled cells (e.g., *see* ([37]), rather than use mass assays. Nevertheless, the durability of this PLC assay may finally be challenged—and by a kit, no less! A homogeneous time resolved fluorescence assay for InsP is now commercially available [38]. The basis of this procedure is that InsP produced by cells competes with an InsP analogue that is conjugated to a fluorophore acceptor, which in turn binds to an anti-InsP monoclonal antibody. Since it is a competition assay, an increase in InsP is recorded as a decrease in fluorescence.

As for the aforementioned discovery of Ins(1,3,4,5,6)P_5 in erythrocytes of birds and turtles, that development owed much to the introduction of nuclear magnetic resonance (NMR) spectroscopy to the analysis of inositol phosphate structure [25]. However, it was not until 1980 that the functional significance of this Ins (1,3,4,5,6)P_5 was found to lie in its regulation of hemoglobin's oxygen affinity [39]; air-breathing fish were subsequently added to this elite group of animals [39]. Changes in their blood oxygen affinity appear to correlate with fluctuations in the levels of Ins (1,3,4,5,6)P_5 within the erythrocyte. That concept represents the first time that changes in the turnover of an inositol phosphate has been considered to reflect a *regulatory* function. Despite that conclusion, this area of animal physiology ***still*** did not receive much attention, perhaps because it was considered merely to be a niche phenomenon. Contemporary interest in the regulation of hemoglobin's oxygen affinity was more generally focused on 2,3-diphosphoglycerate [39]. Nor did anyone seem to wonder whether inositol phosphates might be present in other animal cells, with one isolated and intriguing exception: in 1975, Lapan [40] identified high amounts of $InsP_6$ deposits in the two apical cells of the 28-cell mesozoan dispersal larvae, *Dicyema typus*. Since the $InsP_6$ accounted for almost 50% of the benthic organism's body weight, Lapan suggested that the inositol phosphate served to impart negative buoyancy, although he wisely recognized this was unlikely to be its only function [40].

Following the introduction of NMR to determine inositol phosphate structure (*see* above), this methodology has continued to make significant contributions to the field, although the equipment is complex, expensive, and it requires considerable quantities of the material to be analyzed (again, not a "relatively easy" technique). Despite those drawbacks, NMR would eventually lead to the identification of $InsP_6$ as a cofactor for stimulating a multienzyme complex that repairs DNA through nonhomologous end-joining [41]. NMR has also been deployed to characterize the structures of some of the other inositol phosphates that accumulate in animals [42], slime molds [43, 44] and soils [45]. The existence of inositol phosphates in soils might not seem that surprising, given their prevalence in plants. However, there is a twist in that tale. Most eukaryotes only synthesize phosphorylated versions of the *myo*-inositol scaffold [46, 47]. Yet, in soils, there are not only *myo*-inositol phosphates, but also significant quantities of *scyllo*-inositol phosphates, plus smaller but no less surprising amounts of *chiro*- and *neo*- versions [45]. Amazingly, we still do not know why—see Chapter 3.

3 Inositol Lipids as Sources of Inositol Phosphates: A Concept that Developed Slowly

At a time animal biologists took little interest in inositol phosphates per se, the inositol moiety itself began to gain a different source of attention when it was discovered to be a component of a minor phospholipid—PtdIns. Following the original identification of PtdIns in soybean [48], Folch's group went on to report its presence in cow brain [49]. A few years later Folch described PtdIns4P [50].

Folch did not use any chromatographic techniques to isolate these inositol lipids. Instead, they were purified simply by precipitation and resolubilization in various solvents. While in one sense this may be viewed as a remarkable accomplishment, it is, perhaps, *unremarkable* that his PtdIns4P preparation was later found to be "contaminated" with $PtdInsP_2$; that determination was made in the early 1960s by Clinton Ballou's group [51], and independently by the laboratory of Rex Dawson [52, 53]. By adhering closely to Folch's protocol, Ballou [51] has felt obliged to acknowledge that their work was a substantial undertaking. His fearless technician, Stephen Freer, purified one preparation of $PtdInsP_2$ from a total of 100 cow brains! I estimate that would have required well over 100 liters of acetone for the initial protein precipitation step. Apparently (*see* [54]), the colossal nature of this undertaking prompted Freer to work on the roof of his building in Berkeley California, arguably with an unhealthy disregard for acetone's low flash point. Fortunately, the preparation was rescued from an untimely cremation, and the fire was extinguished, without any (recorded) injury to Freer [54]. In contrast, Rex Dawson worked at the Agricultural Research Council Institute of Animal Physiology in Cambridge, UK, which enjoys a climate that is not so accommodating to outdoor research. Or, perhaps, the Safety Office was more vigilant. In any case, Dawson [52] modified the methodology—he modestly described the revised protocol as "a convenient method"—that scaled down the process to the extent that even guinea-pig brains were successfully used as starting material. Clearly, Dawson's technical improvements gave inositol phospholipid research wider accessibility—as well as making this a less dangerous field of research.

These studies with inositol lipids progressed with little thought to the possibility that the inositol phosphate headgroup might be functional. Nevertheless, it is instructive to continue with the history of this topic, particularly to see how it drove the development of techniques that would later prove to be invaluable for inositol phosphate research. For example, it was lipid-centric papers [55, 56] in which chemical procedures were described that conclusively confirmed the headgroup of $PtdInsP_2$ as being $Ins(1,4,5)P_3$. In those studies, the inositol lipid was chemically degraded to liberate free $Ins(1,4,5)P_3$, which was then subjected to periodate oxidation, which cleaves the ring between two vicinal hydroxyl groups. Afterward, following enzymatic dephosphorylation, and resolution by paper chromatography and electrophoresis, the identification of the resultant polyol as iditol is diagnostic of the nature of the parent inositol phosphate. These procedures would serve important roles in later years for characterizing the structures of many inositol phosphates (*see* below).

The first paper to indicate that inositol lipids might be of particular interest because of their unusually active turnover was published by Rex Dawson in 1954 [57]. He radiolabeled a guinea-pig brain "dispersion" with [^{32}P], and then used alkaline hydrolysis to release glycerophosphoesters, which he resolved by two-dimensional paper chromatography. The parent phospholipid with the highest specific radioactivity—by far—was clearly an inositol lipid [57]. This study attracted the attention of Lowell and Mabel Hokin, who had originally thought acetylcholine-dependent stimulation of [^{32}P]-labeling of the pancreas was due to increased RNA synthesis [58]. Using procedures similar to Dawson's, the Hokins subsequently identified PtdIns as being the major inositol lipid to become radiolabeled with [^{32}P] after addition of a muscarinic agonist [59]. A couple of years later—in 1958—the Hokins reiterated their conclusion following experiments with tissue labeled with [^{3}H]inositol [60]. This biological response—evidence of stimulus-dependent enhancement of the rate of *synthesis* of the inositol lipids—became known as their "PI [phosphatidylinositol] effect" [61, 62].

The 1958 study by the Hokins [60] is also technically noteworthy for it containing a description of the preparation and utilization of [^{3}H]inositol for studying inositol lipid turnover; during the same year, Bernard Agranoff also produced [^{3}H]inositol [63]. Both groups synthesized [^{3}H]inositol by platinum-catalyzed reduction of scyllo-inosose; the Hokins [60] report the heroic consumption of 200 mCi [^{3}H]! (The methodology had originally been developed for the synthesis of deuterium-labeled inositol, first by Stetten and Stetten [64]; later it was refined by Theodore Posternak [47]).

The value of this [^{3}H]inositol synthesis lives on because, in *most* cultured cells of animal origin, the only water-soluble, polar metabolites that accumulate the [^{3}H] are glycerophosphoinositol and the inositol phosphates; this metabolic (and chemical) stability underpins a wealth of inositol phosphate literature. Nevertheless, there is an important exception to inositol's metabolic stability: kidney cells contain a myo-inositol oxygenase, opening up the opportunity for the [^{3}H] in the inositol to be removed and incorporated into nucleotides such as ATP and GTP, which are difficult to resolve from $InsP_3$ [65, 66]; a similar outcome results from in vivo labeling of ascites tumor cells, or whole animal labeling [65]. On the other hand, [^{32}P]-labeling is even less useful for the analysis of cellular inositol phosphates, as that radiolabel is inevitably incorporated into many other water-soluble molecules. A case in point is a study with [^{32}P]-labeled platelets [67] in which the [^{32}P]$InsP_3$ fraction was later shown to be heavily contaminated with 2,3-[^{32}P]diphosphoglycerate [68].

Even though it was appreciated at an early stage that InsP was a potential product of receptor-dependent PtdIns turnover (*see*, for example, [33, 62]), only very occasionally [69] was it tentatively suggested that release of the inositol phosphate headgroup might yield a molecule with its own, separate function, and the idea failed to catch on.

In 1964, Thompson and Dawson [70] suggested that PtdIns hydrolysis was not the only source of cellular InsP. They raised the possibility that InsP may also formed by dephosphorylation of Ins(1,4,5)P_3 and/or Ins(1,4)P_2, in turn produced by PLC-mediated turnover of PtdIns(4,5)P_2 and PtdInsP respectively. Thompson and Dawson obtained data in support of this idea by following enzymatic hydrolysis of PtdIns(4,5)P_2 by an acetone extract of ox brain; they used ice-cold trichloroacetic acid to isolate a water soluble fraction, which were then neutralized by repeated washing in diethyl ether, prior to the electrophoretic separation and identification as $InsP_3$ and/or $InsP_2$.

Remarkably, Thompson and Dawson [70] did not use any radiolabel in their experiments, instead relying on phosphate mass assays. A few years later, Durell and colleagues [71, 72] revisited this topic, and essentially confirmed the earlier results. They labeled inositol lipids from the brain, either in vitro with [^{32}P]ATP or in vivo with [^{3}H]inositol (0.25 mCi injected directly into the cisterna magna!). Acetylcholine was added to various crude membrane fractions (which we now know must have been contaminated with plasma membranes). To isolate enzymatically produced inositol phosphates, ice-cold perchloric acid was preferred over the trichloroacetic acid used by Thompson and Dawson [70]. Durell and colleagues [71, 72] recognized that the addition of lithium hydroxide readily removed the perchlorate and neutralized the solution, thereby avoiding arduous repeated washes with ether, as well as the danger of its flammability. Yet, strangely, the trichloroacetic acid-extraction/ether neutralization procedure was frequently resurrected in later years (e.g., [73]). These days, perchloric acid extraction has become routine for inositol phosphate assays (with potassium carbonate as the more usual neutralizing agent [37]).

Even though Durell and colleagues [71, 72] produced data consistent with receptor-mediated hydrolysis of PtdIns(4,5)P_2 to Ins(1,4,5)P_3, this observation did not initially attract much interest—even from the authors themselves. At that time, there was a consensus of opinion that PLC preferred PtdIns as a substrate over PtdIns(4,5)P_2; for example, *see* [74]. Also, only relatively tiny amounts of Ins(1,4,5)P_3 had been detected. The much higher levels of cellular InsP were more readily quantified. It is almost as if there was an inherent bias that the more abundant of a family of molecules were presumed to be the more functionally significant. Perhaps that may seem an immoderate statement, but I sensed a similar prejudice after we [75] and others [76] first reported the presence of small quantities of inositol pyrophosphates in mammalian cells; a well-respected colleague from a local university enquired as to why I did not prefer to focus my attention on $InsP_6$, since *that* was present in cells at much larger concentrations.

Considering the length of time that it took the field to accept there is functional significance to PLC-dependent production of an inositol phosphate, it is ironic that this concept first gained impetus from a false lead [77, 78]. The culprit was 1,2-cyclic InsP which, in vitro, PLC can be persuaded to produce by hydrolysis of PtdIns [77–79]. Moreover, mammalian cells hydrolyze 1,2-cyclic InsP to Ins1P [80]. The analogy with cAMP production and degradation was obvious. However, after a short-lived flurry of interest in this cyclic inositol phosphate, it was established to be a slowly metabolized, degradation product of the 1,2-cyclic version of Ins(1,4,5)P_3 [81], which in turn is a minor and functionally insignificant side product of PLC-dependent PtdIns(4,5)P_2 hydrolysis [82]: an illustration that evolution is not the route to a perfect enzyme. These cyclic molecules no longer receive any attention, and the 1,2-cyclic InsP phosphohydrolase is considered merely to be a "mopping-up" enzyme.

Research into InsP production took a different, and striking new direction when Allison and Stewart [83] began to explore the actions of lithium, a drug widely used to treat manic depressive behavior. By using mass spectrometry, it was found that the administration of lithium to rats decreased inositol levels in the cerebral cortex [83], and subsequently it was demonstrated that InsP levels reciprocally increased—and in an atropine-sensitive manner [84]. It was separately demonstrated that lithium is an inhibitor of InsP monophosphatase activity [85]; in the presence of lithium, the InsP product was metabolically "trapped" [84, 86]. This lithium-trap technique would subsequently play another vital role in inositol phosphate research (*see* below). Furthermore, mass spectrometry methodology has since been adapted to the assay of inositol polyphosphates, and has proved to be particularly useful in circumstances in which [^{3}H]-radiolabeling radiolabeling is not practical (*see* Chapter 2 provided by Ronald Pegg's group).

This work with lithium heightened interest in the apparent receptor-dependent, PLC-mediated hydrolysis of PtdIns to InsP. Nevertheless, Ins1P, the presumed product of PtdIns hydrolysis, was already known not to be the only InsP isomer in the brain. In addition, Ins3P is formed by cyclization of glucose-6-phosphate [87]. Due to the plane of symmetry across the inositol ring, Ins1P and Ins3P are enantiomers. It is James Allison who has been credited with developing and implementing the resolution of Ins1P and Ins3P by a chiral capillary column coupled to a chemical ionization gas chromatograph-mass spectrometer [88]. It turned out that both InsPs were elevated in the brain by lithium treatment, although the effect upon Ins1P was much larger [88]. The latter authors concluded that Ins1P was likely a product of stimulus-dependent inositol lipid turnover, and, moreover, that the therapeutic effects of lithium might reflect its perturbation of a role of inositol lipid turnover in synaptic function. A sad footnote to that story is that Allison did not live to see the work published; he passed away 2 years prior to publication of this study [88].

4 A Cell-Signaling Role for Ins(1,4,5)P_3

Bob Michell's [89] seminal treatise, published in 1975, drew significantly more attention to the "PI-effect." Not all of the ensuing debate was complimentary; Lowell Hokin once wrote that there were those who were prepared to state that it was all "hokum" and "folly" [90]. Nevertheless, those with a more open mind to new concepts were persuaded by Michell's hypothesis that receptor-mediated PtdIns turnover is intimately linked to Ca^{2+} signaling. Indeed, it was this proposal for a novel signaling event that attracted Michael Berridge to this field of work (*see* [1]). One of his initial goals was to set up an assay of stimulus-dependent InsP accumulation as a readout for PtdIns turnover (as Berridge himself has explained [1]). He used [^{3}H]inositol-labeled blowfly salivary glands as his first model system. The [^{3}H]InsP was assayed by gravity-fed, anion-exchange columns, using the procedures described earlier by Hawthorne's group [33, 35, 74]. Berridge also took note of papers describing the lithium trap: the inhibition of InsP hydrolysis by the inositol monophosphatase (*see* above). This substantially increased the accumulation of InsP in Berridge's experiments [91].

Berridge's early work in this field also coincided with an important change in its focus. Using [^{3}H]inositol labeling of iris muscle, Abdel-Latif and Akhtar [92] had expanded on the aforementioned observation of Durell and Garland in 1969 [71], by demonstrating acetylcholine-activated PtdIns(4,5)P_2 hydrolysis to Ins(1,4,5)P_3. Around the same time, Michell and colleagues were also drawing attention to the possibility that stimulus-activated PtdIns(4,5)P_2 turnover is a primary signaling event to which enhanced PtdIns turnover was only secondary [93–95].

In view of these developments, Berridge and colleagues naturally became intrigued that a substantial receptor-dependent increase in the [^{3}H]InsP fraction was not the only key result to emerge from their experiments with blowfly salivary glands [91]. They observed additional, more polar [^{3}H]inositol-labeled materials [91]. Consequently, they expanded their model systems to include [^{3}H]inositol-prelabeled rat parotid glands and brain slices, and with the help of [^{32}P]-labeled standards prepared from erythrocyte ghosts, they confirmed stimulus-dependent accumulation of [^{3}H]InsP_2 and [^{3}H]InsP_3 [96]. Shortly thereafter, it would emerge that PtdIns(4,5)P_2 and not PtdIns is the main substrate for PLC in vivo; this dramatic reversal of a decades-long consensus also gained strength from an essentially quite simple methodological concept: the performance of assays in vitro that use physiologically relevant conditions [97]! In the latter study, PLC was presented with PtdIns(4,5)P_2 substrate in a lipid mixture with similar composition to that of the inner half of the lipid bilayer of a rat liver plasma membrane, all suspended in a medium with an intracellular-like ionic composition.

Berridge next wanted to determine if Ins(1,4,5)P_3 would control Ca^{2+} mobilization, but he was faced with significant technical hurdles. One issue was to secure a source of Ins(1,4,5)P_3; at the time, it was not commercially available. It was Robin Irvine who came to the rescue, by scaling up the preparations that could be obtained from erythrocyte ghosts, through Ca^{2+}-dependent activation of PLC-mediated PtdIns(4,5)P_2 hydrolysis, as described earlier by Pete Downes and Bob Michell [98]. Berridge also learned that Irene Schulz had developed a saponin-permeabilized pancreatic cell preparation, specifically for the purpose of recording intracellular Ca^{2+} fluxes [99]. The interaction of saponin with cholesterol forms approximately 10 nm holes in the plasma membrane, but it leaves internal membranes intact because of their lower cholesterol content [99]. Berridge realized that the same holes would admit exogenously added Ins(1,4,5)P_3 into the cells. This was the model system used to demonstrate that Ins(1,4,5)P_3 mobilizes cellular Ca^{2+} stores from endoplasmic reticulum [100]. That classic 1983 publication established an entirely new field of research: signal transduction by inositol phosphates. Irvine was also able to provide Ins(1,4,5)P_3 to a number of different laboratories, which was key to the immediate confirmation and wider recognition of this new signaling paradigm. For the first time in the history of inositol phosphate research, it was a hectic, rapidly developing field.

5 Inositol Phosphate Proliferation

The discovery that Ins(1,4,5)P_3 is a cellular signal prompted my own entry into inositol phosphate research, through a contribution to a study into the dephosphorylation and hence inactivation of the Ins(1,4,5)P_3 signal [101]. If that had been the sole pathway of Ins (1,4,5)P_3 metabolism, then almost certainly I would not now be writing this chapter. Fortunately (for my career), inositol phosphate research turned out not to be that basic. For example, even as multiple laboratories were confirming in various cell models both stimulus-dependent PtdIns(4,5)P_2 hydrolysis and Ca^{2+}-mobilization by Ins(1,4,5)P_3, Berridge and colleagues were learning that much of the [^{3}H]InsP_3 accumulation was not actually due to Ins (1,4,5)P_3. Irvine [102] has described how Pete Downes first made this observation in 1982, when he farsightedly introduced high-pressure liquid chromatography to the analysis of inositol phosphate metabolism: a genuine [^{32}P]Ins(1,4,5)P_3 standard was mixed with [^{3}H]InsP_3 produced by carbachol-stimulated parotid glands, and applied to an anion-exchange MonoQ column connected to an FPLC. The two differently radiolabeled versions of the same compound steadfastly refused to coelute precisely [103]. This was a technically significant result for the field, because it demonstrated the importance of routinely separating *isomers* of InsP_3. Consequently, it was not long before Downes' original high-pressure separation technique was modified to be compatible with strong-anion exchange HPLC, courtesy of tetramethylammonium as the fixed functional group; a phosphate-based mobile phase was also developed [104]. Concerns were subsequently raised that these mildly acidic HPLC elution conditions lead to degradation of the pyrophosphorylated members of the inositol phosphate signaling family [105, 106]. However, we have shown this is not an issue [107]. So, this procedure remains the basis for most published HPLC separations of inositol phosphates. In Chapter 7, John York's group expands on the continuing importance of this methodology.

Irvine and colleagues described the structure of the "new" InsP_3 as Ins(1,3,4)P_3 in 1984 [103]. Most of the techniques they deployed were only slight modifications (mainly for safety reasons [102]) of the same chemical and chromatographic techniques that Ballou [51, 55] had described 20 years earlier (another example of methodological longevity in this field). Nevertheless, Len Stephens and colleagues [108] subsequently demonstrated how an HPLC-based resolution of polyols is a more efficient substitute for paper chromatography and electrophoresis. These methodologies would greatly facilitate the characterization of the wide range of inositol phosphates in avian erythrocytes [108, 109], mammalian cells [108, 110, 111], plants [112], *Dictyostelium* [113] and insect cells [114].

Up until 1984, most studies into cellular inositol phosphate metabolism involved incubating cells or tissue slices with [^{3}H] inositol for just a few hours, which is sufficient to detect [^{3}H] $InsP_3$ and [^{3}H]$InsP_4$. But it takes much longer for the more highly phosphorylated inositol phosphates to attain the isotopic equilibrium necessary to study accurately their turnover [115]. (Our rule of thumb—this only works with dividing cells—is to assume that upon cell division, half of the $InsP_6$ in the daughter cells must have undergone de novo synthesis using [^{3}H]Ins. Thus, by 3 or 4 divisions, the $InsP_6$ pool must be at 88% and 94% of isotopic equilibrium, respectively). Indeed, it was the result of an 86 h labeling of GH_4C_1 cells, published in 1985, that finally revealed the existence in animal cells of $InsP_5$ and $InsP_6$ [73]. Suddenly, both plant and animal biologists studying inositol phosphate metabolism had come to appreciate that they were rowing in the same boat.

Not surprisingly, the rather mundane ("housekeeping") idea that $InsP_6$ is a phosphate-storage depot in plants was initially thought to be transferable to animal cells [116]. But eventually, as evidence gathered of the multiple signaling properties of $InsP_6$ and other members of the inositol phosphate family in animals, these ideas crossed back into plant research too [117]. With the key exception for higher plants that Ins(1,4,5)P_3 does not mobilize Ca^{2+} from intracellular stores [117].

As the inositol phosphate community further expanded, an important technical advance came in the application of discerning model eukaryotic systems. For example, rapidly dividing yeast cells can be labeled to equilibrium overnight, which assisted the first demonstrations of stimulus-dependent Ins(1,4,5)P_3 production in yeast cells [118, 119]. The genetic tractability of yeast also did not go unnoticed; it was John York's groundbreaking studies with *Saccharomyces cerevisiae* that enabled him to clone the two genes that convert Ins(1,4,5)P_3 to $InsP_6$, and it was this work that also led to the determination that $InsP_6$ stimulates mRNA export from the nucleus [120, 121].

The proliferation of the inositol phosphate family underpins the requirement for HPLC to analyze inositol phosphates in [^{3}H]-labeled, cultured cells. Unfortunately, the relative expense of [^{3}H] inositol is an impediment ([^{14}C]-inositol is even more costly). The HPLC equipment is also expensive to both purchase and maintain. The volumes of scintillation fluid that are required add further strain to the average laboratory budget. Another downside is the technique's low throughput nature. Nevertheless, [^{3}H]inositol-labeling and HPLC analysis of cellular inositol phosphates has endured because of its sensitivity. The development of routine, alternative technology is constrained by inositol phosphates being unable to adsorb or emit UV or visible wavelength light, and so they cannot be detected by spectrophotometric or fluorescence methods. Another challenge for any mass-based analysis of most inositol phosphates is their low micromolar levels (0.1 to 2 μM) in cultured cells. This requires a minimum sensitivity range of 1-20 pmol (assuming 10^7 cells in a 10 cm dish with a median cell volume of 1 pl [122]).

One alternative mass assay that has enjoyed some useful applications is Georg Mayr's [123] nonradioactive, metal-binding dye detection method. However, the minimum sensitivity is >20 pmol [124], the methodology for sample preparation is rather complicated, and it can only semiquantitatively distinguish between inositol phosphate isomers [125]. The restriction on sensitivity is similar for methods based on chemically suppressed conductivity (>20 pmol [126]) and mass spectrometry (25 pmol; *see* Ronald Pegg's contribution in the current volume); another limitation of the latter method is that it cannot distinguish between inositol phosphate isomers. Nevertheless, these are methods that must be employed in circumstances where radiolabeling is not practical, and/or when environmental samples must be analyzed—the importance of the latter is described in Chapter 3.

An alternative mass-assay method, at least for studying cellular $InsP_6$ and inositol pyrophosphates, is the SDS-PAGE separation and toluidine blue staining protocol introduced by Adolfo Saiardi and colleagues [106], coupled with their development of a procedure that uses TiO_2 beads to purify and concentrate inositol phosphates from cell-free extracts [127]. Studies into the metabolism and function of 5-$InsP_7$ have been made more generally accessible by these new methods, because they are simple and in expensive. Nevertheless, the trade-off is that the dynamic range of the assay and its degree of accuracy are both rather limited. As for $InsP_8$, its cellular concentration is generally below this method's level of detection [107], which is unfortunate because stimulus-dependent changes in cellular levels of $InsP_8$ are a prevalent signaling paradigm in mammalian cells [128–130]. On the other hand, this technique is particularly useful for studying the turnover of the exceptionally high levels of inositol pyrophosphates in slime molds (*see* Chapter 5).

More recently, fluorescent sensors have been developed for use in intact cells, specifically to detect either Ins(1,4,5)P_3 [131] or Ins (1,3,4,5)P_4 [132]. Updates on these inositol phosphate assays are provided in Chapters 4 and 12 contributed by Takashi Morii and Akihiko Tanimura. Enzymatic assays of specific inositol phosphates can also play an important role: an IP6K-based assay for $InsP_6$ (Fig. 1), with nanomolar sensitivity, has established that this polyphosphate is not a constituent of human serum or urine [133]—isolated studies that state the contrary view [134, 135] have been criticized on methodological grounds [127, 136]. This controversy is not simply an esoteric academic argument; it is a topic that impacts the validity of some of the claimed therapeutic benefits of clinical administration of $InsP_6$ [137].

As an increasing number of inositol phosphates have been proposed to have signaling functions, so synthetic chemists have stepped up to produce these molecules for experimental testing of these new ideas. The technical efforts involved (*see* [138]) deserve recognition for advancing research in this field, particularly with regards to the chemical synthesis of inositol pyrophosphates; their especially concentrated array of phosphate groups places further demands upon sample preparation and handling [139]. Thus, the publication of methods for the synthesis of $InsP_7$ isomers [139, 140], metabolically stable analogues in which the diphosphate groups are replaced with phosphonoacetate or methylenebisphosphonates [141–143], and 1,5-$InsP_8$ [144], are all substantial accomplishments. It is unfortunate that the corresponding [^{3}H]-labeled inositol pyrophosphates are not so readily available, but they can be prepared enzymatically from the commercially available [^{3}H]$InsP_6$ precursor [145]. Other methods for inositol pyrophosphate research are described elsewhere in this book: Chapter 6 describes their application to the assay of the DIPPs (Fig. 1). Chapters 11 and 13, one by Sung Hyun Kim, and another by Mi Sook Sung and Tae-Young Yoon, show how to study regulation of synaptic vesicle fusion by $InsP_7$. Finally, Chapter 8 describes how to use a back-phosphorylation assay to study protein pyrophosphorylation by inositol pyrophosphates.

During the 1980s, [^{3}H]Ins(1,4,5)P_3 become available commercially; this was an important boost for the field. For example, it helped Irvine and colleagues discover Ins(1,4,5)P_3 3-kinase activity [146]. In a separate development, Solomon Snyder's group used this radiolabeled material to purify an [^{3}H]Ins(1,4,5)P_3-binding protein from rat cerebellum membranes; this 260,000 kDa protein was incorporated into lipid vesicles and was shown to act as an Ins (1,4,5)P_3-regulated Ca^{2+} channel [147]. A protein of similar size had been independently purified by Katshuiko Mikoshiba's group, but not through any initial interest in inositol phosphates. Their motive was research into information processing by Purkinje cells, in which their protein—P_{400}—was particularly enriched [148]. Snyder's studies helped Mikoshiba's group realize that their P_{400} was the Ins(1,4,5)P_3 receptor [148]. This work naturally attracted the attention of synthetic chemists, and informative inositol phosphate analogues were developed, yielding, for example, new insight into how ligand binding to the Ins(1,4,5)P_3 receptor is transduced into a conformational change that opens the Ca^{2+} pore [149]. Chapter 9 describing methods for studying the Ins(1,4,5)P_3 receptor is contributed by Ana Rossi and Colin Taylor. The development of a nonmetabolizable trisphosphorothiate analogue of Ins(1,4,5)P_3 for functional experiments in cell-free systems was also useful in eliminating possible activities of Ins(1,4,5)P_3 metabolites from those of Ins(1,4,5)P_3 itself [150].

The enormous success that ultimately came from using radiolabeled Ins(1,4,5)P_3 to purify its functional receptor inevitably led to efforts to gain similar rewards for other members of this signaling family, most notably Ins(1,3,4,5)P_4, and InsP_6. Probes that were either radiolabeled [151] or photoaffinity-labeled [152] were deployed. Additionally, Glenn Prestwich was the first to immobilize Ins(1,3,4,5)P_4 for identification of its potential binding proteins [153]. With regards to Ins(1,3,4,5)P_4, the resulting purification, cloning and functional characterization of one particular "Ins(1,3,4,5)P_4-binding protein," so-called GAP1^{IP4BP} [154], is typical of the general outcome: this application for labeled Ins (1,3,4,5)P_4—the headgroup of the inositol lipid PtdIns(3,4,5) P_3—has led to the identification and characterization of proteins for which the *lipid* is now generally considered to be the functional ligand [155]. The latter statement should not be taken as demeaning the accomplishment, it simply reflects my opinion that this type of work mainly benefitted research into inositol lipids, not inositol phosphates. Nevertheless, competition between PtdIns(3,4,5)P_3 and Ins(1,3,4,5)P_4 may be functionally significant in some circumstances—*see*, for example [156]. But a standalone function for any Ins(1,3,4,5)P_4-specific "receptor" remains to be demonstrated.

As for InsP_6 "probes," an early use of the [^{3}H]-labeled material led to purification and sequencing of a putative "InsP_6-receptor": the AP-2 clathrin assembly protein [157]. However—and this may seem to be a recurring theme—the physiologically relevant ligand is now generally considered to be an inositol lipid, in this case, PtdIns (4,5)P_2 [158]. This outcome is illustrative of a technical difficulty inherent in the study of InsP_6: in vitro at least, it is *so* polar that it can bind nonspecifically to electropositive protein domains through electrostatic interactions that are delocalized [159], that is, "brute-force" in nature [3]. This phenomenon can play havoc with a search for genuine, InsP_6-specific binding proteins. Another technical issue is that InsP_6-binding data are also sensitive to the components of the assay buffer, and in particular the nature of the counter-ions for the polyphosphate [160, 161]. A pentamagnesium salt of InsP_6 is recommended for experimental purposes since it is likely the most physiological, according to a high resolution potentiometric study of its solution equilibria [161].

There is another property of InsP_6 that, so far, has defied explanation: it will elute from a gel filtration column with an apparent mass of 9 kDa—around nine times what it actually is [162]. InsP_6 is also not readily dialyzable [163]. The prospect that InsP_6 may somehow aggregate into a higher-order macromolecular complex is mind-boggling, be it either as an artifact in vitro or a quite natural occurrence in vivo.

Another InsP_6 probe that proved useful was 2-(6-aminohexyl-1-phospho-InsP_5 attached to Affi-Gel resin (an "InsP_6 affinity column"). This was used to purify the Ins(1,3,4)P_3 6-kinase

[164]. Years later, an alternative $InsP_6$ affinity column was developed: the 2-position was attached to an aminohexyl linker to biotin, which was immobilized on streptavidin beads [165]. Among the proteins that were isolated using this resin were several that participate in nonhomologous end joining of DNA, confirming an earlier report that $InsP_6$ is a cofactor for this form of DNA repair [41]. More recently, yeast cell lysates were screened for proteins that would bind to a bead-immobilized, metabolically stable analogue of 5-$InsP_7$ [166]. The bound proteins were subjected to gene ontology analysis, thereby revealing a significant over-representation of proteins with bioenergetically important roles, such as nucleotide metabolism, glucose metabolism and ribosome biogenesis [166]. However, it should be noted that there is always the possibility that the affinity of an inositol phosphate for a target protein may be compromised by the carbon to which the linker is attached. Additionally, proteins identified in such studies may not bind directly to the immobilized bait, but to other proteins trapped on the resin.

6 Ongoing Challenges

One insightful approach to study inositol phosphate function can be to specifically elevate its concentration inside intact cells. However, the highly polar nature of inositol phosphates prevents them from crossing the plasma membrane. This issue can be circumvented by microinjection, in experiments that utilize single cells—as in the patch clamp set-up, for example. However, an alternative approach is required when populations of intact cells are to be studied. This was the reason for the development of membrane-permeant analogues that can be bioactivated once inside cells.

The earliest efforts in this area involved masking the hydroxyl and phosphate groups of Ins(3,4,5,6)P_4 with butyrate, and pairs of acetylmethoxy-esters, respectively [167]. This stealth version of Ins(3,4,5,6)P_4 is able to diffuse into cultured cells; intracellular esterases unmask the native inositol phosphate. This particular technical advance was a key step in the discovery that Ins(3,4,5,6)P_4 is a highly specific inhibitor of a specific Ca^{2+}-regulated Cl^- current in the plasma membrane [167]. Nevertheless, this is an approach that only succeeds if the rate of metabolism of the free inositol phosphate is slower than its rate of generation from the cell-permeant analogue. This is indeed the case for Ins(3,4,5,6)P_4; its metabolism is relatively slow [110], and so upon dosage with the cell-permeant version, free Ins(3,4,5,6)P_4 accumulates inside cells despite the time it takes for the ten masking groups to be cleaved [167]. In contrast, there is much faster intracellular metabolism of Ins(1,4,5)P_3 so its cell-permeant analogues are less efficient for signaling research [168]. A cell-permeant caged version of Ins

$(1,4,5)P_3$ appears to be more useful [169]. In Chapter 10, Peter Lipp provides an update on applications for caged Ins$(1,4,5)P_3$.

The introduction into cells of the substantially more polar inositol pyrophosphates has been a greater technical challenge. Recently, however, a caged version of 5-InsP_7 was loaded into HeLa cells using a guanidinium-rich molecular transporter [170]; photouncaging of the 5-InsP_7 confirmed earlier ideas [171] concerning its ability to inhibit receptor-dependent recruitment of AKT to the plasma membrane.

Research into the actions of 5-InsP_7 has generated another technical hurdle to be overcome: how to determine if intact cells can recapitulate its pyrophosphorylation of serine residues that to date has only been observed in vitro [172, 173]. Unfortunately, no-one has so far produced any anti-serine-pyrophosphate antibodies. As previously noted [174], the value of such antibodies would be limited by their epitope recognition being too specific to the amino-acid sequence of the original immunogen.

There are also serious impediments to the identification of pyrophosphorylated peptides by mass spectrometry [174]: the chemical lability of phospho-serine, the possibility of low stoichiometry of the modification, and an inability to distinguish a diphosphate within a peptide sequence from two proximal monophosphates. Currently, the best available method, while valuable, is indirect: in Chapter 8, Rashna Bhandari describes how "back-phosphorylation" of a protein isolated from a cell extract can be used to ascertain its pyrophosphorylation status in vivo.

Another ongoing challenge, particularly in studies of signaling by InsP_6 and the inositol pyrophosphates, is to gain insight into their subcellular compartmentalization. This is an important goal because total cellular levels of InsP_6 are 20-fold higher than those of 5-InsP_7; InsP_8 is even less bountiful [107]. If all of the InsP_6 were to be freely distributed around the cytoplasm, it could be expected to compete successfully for proteins that might mediate actions of the less abundant inositol pyrophosphates. For example, in vitro, InsP_6 can block the regulatory consequences of 5-InsP_7 binding to pleckstrin homology domains [171]. Also in vitro, InsP_6 inhibits protein pyrophosphorylation by 5-InsP_7 [173]. Fortunately, there are good reasons to believe that much of the cellular InsP_6 *is* compartmentalized. For example, a study published in 1993 [175] provides evidence that certain cations form an ionic "bridge" between the phosphates of InsP_6 and anionic phospholipids. That is, intracellular membranes may be wallpapered with InsP_6, thereby lowering its cytoplasmic levels. Another, unknown proportion of intracellular InsP_6 has a nonexchangeable role as a structural cofactor; this is the case for Adenosine Deaminase Acting on RNA type 2 (ADAR2) [176]. The strong suspicion that there are other examples is underscored by PDB entry 4XNH, which describes InsP_6 embedded in the structure of yeast N-terminal acetyltransferase, NatE. Such proteins clearly sequester $InsP_6$ away from the cytoplasm. Just how many others are there?

To interrogate this possibility of $InsP_6$ compartmentalization, it would be very helpful to have probes that can image its free, intracellular pools in real-time. Perhaps "split" ligand-binding domains might prove useful [177]. Such probes are discussed in Chapter 4 written by Sakaguchi et al.

Research into the functions of inositol phosphates in intact cells would benefit from selective, cell-permeant inhibitors of specific inositol phosphate kinases. These would make a useful complement to genetic intervention in PP-InsP signaling processes; a problem with gene deletion as an experimental approach is that it is inherently long-term. There is always the possibility that a particular phenotype may arise out of secondary genetic changes. One observation that is particularly pertinent to this topic is the altered degree of transcription of over 900 genes in *S. cerevisiae* ($\geq$two-fold change in expression), following the deletion of *vip1*, the yeast ortholog of PPIP5K [178]. This raises ample opportunity for off-target, secondary genetic changes to yield confounding data. Such a high degree of genetic penetration reflects functional promiscuity, since most inositol phosphate kinases have scaffolding roles that are separate from their catalytic roles [179–181].

The best current implementation of inositol phosphate kinase pharmacology is the cell-permeant pan-IP6K inhibitor, *N*2-(*m*-(trifluoromethyl)benzyl) *N*6-(*p*-nitrobenzyl)purine (TNP) [182]. TNP competes for the nucleotide binding site, but because IP6Ks (Fig. 1) are highly unusual kinases in that they exhibit low affinity for ATP, the drug is much less effective against other inositol phosphate kinases. However, off-target effects of TNP remain a concern [183], so alternative inhibitors are required. Another important challenge for the future will be to develop inhibitors that can differentiate between the three different IP6Ks.

Of course, IP6K inhibition does not just affect 5-$InsP_7$ levels; $InsP_8$ synthesis is also compromised (Fig. 1). Thus, TNP, by itself, does not allow us to distinguish which of these two inositol pyrophosphates might be functional. This problem worsens when we target kinases that act at earlier steps in the linear inositol phosphate metabolic pathway. For example, attenuation of an $InsP_3$ kinase activity not only reduces $InsP_4$ levels but also impacts the further synthesis of $InsP_5$, $InsP_6$, $InsP_7$, and $InsP_8$! This constrains the value of either genetic or pharmacological intervention in inositol phosphate signaling.

In view of the above challenges, perhaps a case can be made that inositol phosphate research, which is already technically complex, is in many ways becoming even more difficult, rather than easier. It is hard to imagine how any of the current methodological issues mentioned above may be solved by an announcement that "there is a kit for that."

Acknowledgments

Work in the author's laboratory is supported by the Intramural Research Program of the NIH/National Institute of Environmental Health Sciences.

References

1. Berridge MJ (2005) Unlocking the secrets of cell signaling. Annu Rev Physiol 67:1–21
2. Shears SB (2004) How versatile are inositol phosphate kinases? Biochem J 377:265–280
3. Shears SB (2018) Intimate connections: inositol pyrophosphates at the interface of metabolic regulation and cell-signaling. J Cell Physiol 233:1897–1912
4. Majerus PW (1992) Inositol phosphate biochemistry. Annu Rev Biochem 61:225–250
5. Barker CJ, Berggren PO (2013) New horizons in cellular regulation by inositol polyphosphates: insights from the pancreatic Beta-cell. Pharmacol Rev 65(2):641–669
6. Ganapathi SB, Wei S-G, Lamb FS, Shears SB (2013) Functional regulation of Clc-3 in the migration of vascular smooth muscle cells. Hypertension 61:174–179
7. Kim E, Beon J, Lee S, Park J, Kim S (2015) Ipmk: a versatile regulator of nuclear signaling events. Adv Biol Regul 61:25–32
8. Jordan PA, Kayser-Bricker KJ, Miller SJ (2010) Asymmetric phosphorylation through catalytic P(III) phosphoramidite transfer: enantioselective synthesis of D-myo-inositol-6-phosphate. Proc Natl Acad Sci U S A 107 (48):20620–20624
9. Wilcox RA, Safrany ST, Lampe D, Mills SJ, Nahorski SR, Potter BVL (1994) Modification at C2 of *myo*-inositol 1,4,5-trisphosphate produces inositol trisphosphates and tetrakisphosphates with potent biological activities. Eur J Biochem 223:115–124
10. Tan Z, Bruzik KS, Shears SB (1997) Properties of the inositol 3,4,5,6-tetrakisphosphate 1-kinase purified from rat liver. Regulation of enzyme activity by inositol 1,3,4-trisphosphate. J Biol Chem 272:2285–2290
11. Yang X, Shears SB (2000) Multitasking in signal transduction by a promiscuous human ins(3,4,5,6)P4 1-kinase/ins(1,3,4)P3 5/6-kinase. Biochem J 351:551–555
12. Wilson MP, Majerus PW (1996) Isolation of inositol 1,3,4-trisphosphate 5/6-kinase, Cdna cloning, and expression of recombinant enzyme. J Biol Chem 271:11904–11910
13. Scherer J (1850) Uber Eine Neue Aus Dem Muskelfleisch Gewonnene Zuckerart. Liebigs Ann Chem 73:322–322
14. Nagai Y, Kimura Y (1958) Detection of inositol and inositol diphosphate on paper chromatograms. Nature 181(4625):1730–1731
15. Posternak S (1919) Sur la synthése de l'ether hexaphosphorique de l'inosite avec le principe phospho-organique de réserve des plantes vertes. C R Acad Sci 169:138–140
16. Posternak S (1903) Sur un nouveau principe phospho-organique d'origine végétale, la phytine. Compt rend Soc de biol 55:1190–1192
17. Suzuki U, Yoshimura K (1907) Ueber die verbreitung von "anhydro-oxy-methulen-diphosphor-sauren salzen" oder "phytin" in pflanzen. Tokyo Imp Univ, College of Agr Bul 8:495–502
18. Suzuki U, Yoshimura K, Takaishi T (1907) Ueber ein enzym "phytase" das "anhydro-oxy-methylendiphosphorsäure" spaltet. The bulletin of the College of Agriculture, Tokyo Imperial University 7:503–512
19. Heubner W, Stadler S (1914) Über eine titrationmethode zur bestimmung des phytins. Biochem Z 64:422–437
20. McCance RA, Widdowson EM (1935) Phytin in human nutrition. Biochem J 29 (12):2694–2699
21. Abelson PH (1999) A potential phosphate crisis. Science 283:2015
22. Rapoport S (1940) Phytic acid in avian erythrocytes. J Biol Chem 135:403–406
23. Rapoport S, Guest GM (1941) Distribution of acid-soluble phosphorous in blood cells of various vertebrates. JBC 138:269–282
24. Fiske CH, Subbarow Y (1925) The colorimetric determination of phosphorous. J Biol Chem 66:375–400
25. Johnson LF, Tate ME (1969) Structure of "phytic acids". Can J Chem 47:63–73
26. Desjobert A, Petek F (1956) Paper chromatography of inositol phosphates; use in study of hydrolytic degradation of inositol

hexaphosphate. Bull Soc Chim Biol (Paris) 38 (5-6):871–883

27. Anderson G (1955) Paper chromatography of inositol phosphates. Nature 175:863. https://doi.org/10.1038/175863b0
28. Arnold PW (1956) Paper ionophoresis of inositol phosphates, with a note on the acid hydrolysates of phytic acid. Biochim Biophys Acta 19(3):552–554
29. Hanes CS, Isherwood FA (1949) Separation of the phosphoric esters on the filter paper chromatogram. Nature 164 (4183):1107–1112. illust
30. Wade HE, Morgan DM (1953) Detection of phosphate esters on paper chromatograms. Nature 171(4351):529–530
31. Bohm P, Richarz G (1954) Quantitative determination of inositol in phosphatides. Hoppe Seylers Z Physiol Chem 298 (3-5):110–120
32. Cosgrove DJ (1963) The isolation of myoinositol pentaphosphates from hydrolysates of phytic acid. Biochem J 89:172–175
33. Hubscher G, Hawthorne JN (1957) The isolation of inositol monophosphate from liver. Biochem J 67(3):523–527
34. Smith DH, Clark FE (1952) Chromatographic separations of inositol phosphorus compounds. Soil Sci Soc Am Proc 16 (2):170–172
35. Hawthorne JN, Hubscher G (1959) Separation of glycerylphosphoryl inositol and related compounds on ion-exchange columns. Biochem J 71(1):195–200
36. Tomlinson RV, Ballou CE (1962) Myoinositol polyphosphate intermediates in the dephosphorylation of phytic acid by phytase. Biochemistry 1:166–171
37. Shears SB (1997) Measurement of inositol phosphate turnover in intact cells and cell-free systems. In: Shears SB (ed) Signalling by inositides: a practical approach. Oxford University Press, Oxford, pp 33–52
38. Popiolek M, Nguyen DP, Reinhart V, Edgerton JR, Harms J, Lotarski SM, Steyn SJ, Davoren JE, Grimwood S (2016) Inositol phosphate accumulation in vivo provides a measure of muscarinic M1 receptor activation. Biochemistry 55(51):7073–7085
39. Isaacks RE, Harkness DR (1980) Erythrocyte organic phosphates and hemoglobin function in birds reptiles and fishes. Amer Zool 20:115–129
40. Lapan EA (1975) Magnesium inositol Hexaphosphate deposits in Mesozoan dispersal larvae. Exp Cell Res 94:277–282
41. Hanakahi LA, Bartlet-Jones M, Chappell C, Pappin D, West SC (2000) Binding of inositol phosphate to DNA-Pk and stimulation of double-Strand break repair. Cell 102:721–729
42. Lin H, Fridy PC, Ribeiro AA, Choi JH, Barma DK, Vogel G, Falck JR, Shears SB, York JD, Mayr GW (2009) Structural analysis and detection of biological inositol pyrophosphates reveals that the Vip/Ppip5k family are 1/3-kinases. J Biol Chem 284:1863–1872
43. Martin J-B, Foray M-F, Klein G, Satre M (1987) Identification of inositol Hexakisphosphate in 32p-Nmr spectra of Dictyostelium amoebae. Relevance to intracellular Ph determination. Biochim Biophys Acta 931:16–25
44. Laussmann T, Hansen A, Reddy KM, Reddy KK, Falck JR, Vogel G (1998) Diphospho-*myo*-inositol phosphates in *Dictyostelium* and *Polysphondylium*: identification of a new bisdiphospho-*myo*-inositol tetrakisphosphate. febs 426:145–150
45. Turner B, Paphazy M, Haygarth P, McKelvie I (2002) Inositol phosphates in the environment. Philos Trans R Soc Lond [Biol] 357:449–469
46. Posternak T, Schopfer WH, Reymond D (1942) Recherches Dans La Série Des Cyclites Vi. Sut La Configuration De La Méso-Inosite, De La Scyllite Et D'un Inosose Obtenu Par Voie Biochimique (Scyllo-Ms-Inosose). Helv Chim Acta 25:746–752
47. Posternak T, Schopfer WH, Reymond D (1955) Biochimie Des Cyclitols I. Contribution À l'étude Du Métabolisme Du Méso-Inositol Chez Le Rat. Helv Chim Acta 150:1283–1288
48. Klenk E, Sakai R (1939) Inositmonophosphorsäure, Ein Spaltprodukt Der Sojabohnenphosphatide Z physiol Chem 258:33–38
49. Folch J, Wooley DW (1942) Inositol, a constituent of brain phosphatide. J Biol Chem 142:963–964
50. Folch J (1949) Brain Diphosphoninositide, a new phosphatide having inositol Metadiphosphate as a constituent. J Biol Chem 177 (2):505–519
51. Grado C, Ballou CE (1961) Myo-inositol phosphates obtained by alkaline hydrolysis of beef brain Phosphoinositide. J Biol Chem 236:54–60
52. Dittmer JC, Dawson RM (1961) The isolation of a new lipid, triphosphoinositide, and monophosphoinositide from ox brain. Biochem J 81:535–540

53. Dawson RM, Dittmer JC (1961) Evidence for the structure of brain triphosphoinositide from hydrolytic degradation studies. Biochem J 81:540–545
54. Irvine RF (2016) A short history of inositol lipids. J Lipid Res 57(11):1987–1994
55. Tomlinson RV, Ballou CE (1961) Complete characterization of the myo-inositol polyphosphates from beef brain phosphoinositide. J Biol Chem 236:1902–1906
56. Brown DM, Stewart JC (1966) The structure of triphosphoinositide from beef brain. Biochim Biophys Acta 125(3):413–421
57. Dawson RMC (1954) The measurement of 32p labelling of individual Kephalins and lecithin in a small sample of tissue. Biochim Biophys Acta 14:374–379
58. Hokin LE (1952) The role of ribonucleic acids in amylase secretion by pancreatic slices. Biochim Biophys Acta 8: 225–226
59. Hokin LE, Hokin MR (1955) Effects of acetylcholine on the turnover of phosphoryl units in individual phospholipids of pancreas slices and brain cortex slices. Biochim Biophys Acta 18:102–110
60. Hokin LE, Hokin MR (1958) Phosphoinositides and protein secretion in pancreas slices. J Biol Chem 233(4):805–810
61. Hokin L (1969) Functional activity in glands and synaptic tissue and the turnover of phosphatidylinositol. Ann N Y Acad Sci U S A 165:695–709
62. Hokin MR, Hokin LE (1964) Interconversions of phosphatidylinositol and phosphatidic adid involved in the response to acetylcholine in the salt gland. In: Dawson RMC, Rhodes DN (eds) Metabolism and physiological significance of lipids. John Wiley and Sons, London, pp 423–434
63. Agranoff BW, Bradley RM, Brady RO (1958) The enzymatic synthesis of inositol phosphatide. J Biol Chem 233(5):1077–1083
64. Stetten MR, Stetten D Jr (1946) Biological conversion of inositol into glucose. J Biol Chem 164:85–91
65. Christensen SC, Kolbjorn JA, Simonsen LO (2003) Aberrant 3H in Ehrlich mouse ascites tumor cell nucleotides after in vivo labeling with myo-[2-3H]- and L-myo-[1-3H]inositol: implications for measuring inositol phosphate Signaling. Anal Biochem 313 (2):283–291
66. Christensen S, Harbak H, Simonsen LO (1994) Aberrant 3H labelling of Atp during in vivo labelling of Ehrlich mouse ascites tumour cells with [2-3H]inositol is significant in the study of isomers of Insp3 and Insp4. Biochem J 300(Pt 3):859–863
67. Agranoff BW, Murthy P, Seguin EB (1983) Thrombin-induced phosphodiesteratic cleavage of phosphatidylinositol bisphosphate in human platelets. J Biol Chem 258:2076–2078
68. Dangelmaier CA, Daniel JL, Smith JB (1986) Determination of basal and stimulated levels of inositol triphosphate in [32p]orthophosphate-labeled platelets. Anal Biochem 154 (2):414–419
69. Hawthorne JN (1960) The inositol phospholipids. J Lipid Res 1:255–280
70. Thompson W, Dawson RM (1964) The triphosphoinositide phosphodiesterase of brain tissue. Biochem J 91(2):237–243
71. Durell J, Garland JT (1969) Acetylcholine-stimulated phosphodiesteratic cleavage of phosphoinositides: hypothetical role in membrane depolarization. Ann N Y Acad Sci 165 (2):743–754
72. Durell J, Sodd MA, Friedel RO (1968) Acetylcholine stimulation of the phosphodiesteratic cleavage of guinea pig brain phosphoinositides. Life Sci 7:363–368
73. Heslop JP, Irvine RF, Tashjian AH, Berridge MJ (1985) Inositol tetrakis- and pentakisphosphates in Gh_4 cells. J Exp Biol 119:395–401
74. Kemp P, Hubscher G, Hawthorne JN (1961) Phosphoinositides. 3. Enzymic hydrolysis of inositol-containing phospholipids. Biochem J 79:193–200
75. Menniti FS, Miller RN, Putney JW Jr, Shears SB (1993) Turnover of inositol polyphosphate pyrophosphates in pancreatoma cells. J Biol Chem 268:3850–3856
76. Stephens LR, Radenberg T, Thiel U, Vogel G, Khoo K-H, Dell A, Jackson TR, Hawkins PT, Mayr GW (1993) The detection, purification, structural characterization and metabolism of diphosphoinositol pentakisphosphate(S) and bisdiphosphoinositol tetrakisphosphate(S). J Biol Chem 268:4009–4015
77. Lapetina EG, Michell RH (1973) A membrane-bound activity catalysing phosphatidylinositol breakdown to 1,2-diacylglycerol, D-myo-inositol 1:2-cyclic phosphate and D-myoinositol 1-phosphate. Biochem J 131:433–442
78. Michell RH, Lapetina EG (1972) Production of cyclic inositol phosphate in stimulated tissues. Nat New Biol 240(104):258–260
79. Dawson RM, Freinkel N, Jungalwala FB, Clarke N (1971) The enzymic formation of myoinositol 1:2-cyclic phosphate from phosphatidylinositol. Biochem J 122(4):605–607

80. Dawson RMC, Clarke N (1972) D-Myoinositol 1:2-cyclic phosphate 2-phosphohydrolase. Biochem J 127:113–118
81. Connolly TM, Wilson DB, Bross TE, Majerus PW (1986) Isolation and characterization of the inositol cyclic phosphate products of phosphoinositide cleavage by phospholipase C. metabolism in cell-free extracts. J Biol Chem 261:122–126
82. Hughes AR, Takemura H, Putney JW Jr (1988) Kinetics of inositol 145-Trisphosphate and inositol cyclic 1:2,4,5-trisphosphate metabolism in intact rat parotid acinar cells. Relationship to calcium signalling. J Biol Chem 263:10314–10319
83. Allison JH, Stewart MA (1971) Reduced brain inositol in Lithium-treated rats. Nat New Biol 233(43):267–268
84. Allison JH, Blisner ME, Holland WH, Hipps PP, Sherman WR (1976) Increased brain myo-inositol 1-phosphate in lithium-treated rats. Biochem Biophys Res Commun 71 (2):664–670
85. Naccarato WF, Ray RE, Wells WW (1974) Biosynthesis of myo-inositol in rat mammary gland. Isolation and properties of the enzymes. Arch Biochem Biophys 164:194–201
86. Hallcher LM, Sherman WR (1960) The effects of lithium ion and other agents on the activity of myo-inositol-1-phosphatase from bovine brain. J Biol Chem 255:10896–10901
87. Eisenberg F Jr, Bolden AH (1965) D-myo--inositol-1-phosphate, an intermediate in the biosynthesis of inositol in the mammal. Biochem Biophys Res Commun 21(2):100–105
88. Sherman WR, Leavitt AL, Honchar MP, Hallcher LM, Phillips BE (1981) Evidence that lithium alters phosphoinositide metabolism: chronic administration elevates primarily D-myo-inositol-1-phosphate in cerebral cortex of the rat. J Neurochem 36 (6):1947–1951
89. Michell RH (1975) Inositol phospholipids and cell surface receptor function. Biochim Biophys Acta 415:81–147
90. Hokin LE (1987) The road to the phosphoinositide-generated second messengers. Trends Pharmacol Sci 8:53–56
91. Berridge MJ, Downes CP, Hanley MR (1982) Lithium amplifies agonist-dependent phosphatidylinositol responses in brain and salivary glands. Biochem J 206:587–595
92. Akhtar RA, Abdel-Latif AA (1980) Requirement for calcium ions in acetylcholine-stimulated phosphodiesteratic cleavage of phosphatidyl-myo-inositol 4,5-bisphosphate in rabbit iris smooth muscle. Biochem J 192 (3):783–791
93. Kirk CJ, Creba JA, Downes CP, Michell RH (1981) Hormone-stimulated metabolism of inositol lipids and its relationship to hepatic receptor function. Biochem Soc Trans 9:377–379
94. Michell RH, Kirk CJ, Jones LM, Downes CP, Creba JA (1981) The stimulation of inositol lipid metabolism that accompanies calcium mobilization in stimulated cells: defined characteristics and unanswered questions. Philos Trans R Soc Lond [Biol] 296:123–138
95. Creba J, Downes CP, Hawkins PT, Brewster G, Michell RH, Kirk CJ (1983) Rapid breakdown of phosphatidylinositol-4-phosphate and phosphatidylinositol-4,5-bisphosphate in rat hepatocytes stimulated by vasopressin and other Ca^{2+}-mobilising hormones. Biochem J 212:733–747
96. Berridge MJ, Dawson RM, Downes CP, Heslop JP, Irvine RF (1983) Changes in the levels of inositol phosphates after agonist-dependent hydrolysis of membrane phosphoinositides. Biochem J 212:473–482
97. Irvine RF, Letcher AJ, Dawson RM (1984) Phosphatidylinositol-4,5-bisphosphate phosphodiesterase and phosphomonoesterase activities of rat brain. Some properties and possible control mechanisms. Biochem J 218 (1):177–185
98. Downes CP, Mussat MC, Michell RH (1982) The inositol trisphosphate phosphomonoesterase of the human erythrocyte membrane. Biochem J 203:169–177
99. Schulz I, Kimura T, Wakasugi H, Haase W, Kribben A (1981) Analysis of Ca2+ fluxes and Ca2+ pools in pancreatic Acini. Philos Trans R Soc Lond Ser B Biol Sci 296(1080):105–113
100. Streb H, Irvine RF, Berridge MJ, Schulz I (1983) Release of Ca^{2+} from a nonmitochondrial store in pancreatic cells by inositol-1,4,5-trisphosphate. Nature 306:67–68
101. Storey DJ, Shears SB, Kirk CJ, Michell RH (1984) Stepwise enzymic dephosphorylation of inositol 1,4,5-trisphosphate to inositol in liver. Nature 312:374–376
102. Irvine R (2016) A tale of two inositol Trisphosphates. Biochem Soc Trans 44 (1):202–211
103. Irvine RF, Letcher AJ, Lander DJ, Downes CP (1984) Inositol trisphosphates in

Carbachol-stimulated rat parotid glands. Biochem J 223:237–243
104. Irvine RF, Anggard EE, Letcher AJ, Downes CP (1985) Metabolism of inositol 1,4,5-trisphosphate and inositol 1,3,4-trisphosphate in rat parotid glands. Biochem J 229:505–511
105. Brown NW, Marmelstein AM, Fiedler D (2016) Chemical tools for interrogating inositol pyrophosphate structure and function. Chem Soc Rev 45(22):6311–6326
106. Pisani F, Livermore T, Rose G, Chubb JR, Gaspari M, Saiardi A (2014) Analysis of Dictyostelium Discoideum inositol pyrophosphate metabolism by gel electrophoresis. PLoS One 9(1):e85533
107. Gu C, Wilson MSC, Jessen HJ, Saiardi A, Shears SB (2016) Inositol pyrophosphate profiling of two Hct116 cell lines uncovers variation in Insp8 levels. PLoS One 11: e0165286
108. Stephens LR, Hawkins PT, Carter N, Chahwala SB, Morris AJ, Whetton AD, Downes PC (1988) L-myo-inositol 1,4,5,6-tetrakisphosphate is present in both mammalian and avian cells. Biochem J 249:271–282
109. Stephens LR, Hawkins PT, Downes CP (1989) An analysis of myo-[^{3}H]inositol trisphosphate found in myo-[^{3}H]inositol prelabelled avian erythrocytes. Biochem J 262:727–737
110. Menniti FS, Oliver KG, Nogimori K, Obie JF, Shears SB, Putney JW Jr (1990) Origins of *myo*-inositol tetrakisphosphates in agonist-stimulated rat pancreatoma cells stimulation by Bombesin of *myo*-inositol 1,3,4,5,6-pentakisphosphate breakdown to *myo*-inositol 3,4,5,6-tetrakisphosphate. J Biol Chem 265:11167–11176
111. Wong NS, Barker CJ, Morris AJ, Craxton A, Kirk CJ, Michell RH (1992) The inositol phosphates of Wrkl rat mammary tumour cells. Biochem J 286:459–468
112. Brearley CA, Hanke DE (1996) Inositol phosphates in the duckweed Spirodela Polrhiza L. Biochem J 314:215–225
113. Stephens LR, Irvine RF (1990) Stepwise phosphorylation of myo-inositol leading to myo-inositol hexakisphosphate in dictyostelium. Nature 346:580–583
114. Zhou Y, Wu S, Wang H, Hayakawa Y, Bird GS, Shears SB (2012) Activation of Plc by an endogenous cytokine (Gbp) in drosophila S3 cells and its application as a model for studying inositol phosphate signalling through Itpk1. Biochem J 448(2):273–283
115. Wilson MS, Saiardi A (2017) Importance of radioactive labelling to elucidate inositol polyphosphate signalling. Top Curr Chem (J) 375 (1):14
116. Berridge MJ, Irvine RF (1984) Inositol trisphosphate, a novel second messenger in cellular signal transduction. Nature 312:315–321
117. Gillaspy GE (2011) The cellular language of myo-inositol signaling. New Phytol 192 (4):823–839
118. Schomerus C, Küntzel H (1992) Cdc25-dependent induction of inositol 1,4,5-trisphosphate and diacylglycerol in Saccharomyces Cerevisiae by nitrogen. FEBS Lett 307:249–252
119. Ongusaha PP, Hughes PJ, Davey J, Michell RH (1998) Inositol hexakisphosphate in *Schizosaccharomyces Pombe*: synthesis from ins (1,4,5)P_3 and osmotic regulation. BJ 335:671–679
120. Odom AR, Stahlberg A, Wente SR, York JD (2000) A role for nuclear inositol 1,4,5-trisphosphate kinase in transcriptional control. Science 287:2026–2029
121. York JD, Odom AR, Murphy R, Ives EB, Wente SR (1999) A phospholipase C-dependent inositol polyphosphate kinase pathway required for efficient messenger Rna export. Science 285:96–100
122. Tzur A, Kafri R, LeBleu VS, Lahav G, Kirschner MW (2009) Cell growth and size homeostasis in proliferating animal cells. Science 325 (5937):167–171
123. Mayr GW (1988) A novel metal-dye detection system permits picomolar-range H.P.L.C. analysis of inositol polyphosphates from non-radioactively labelled cell or tissue specimens. Biochem J 254:585–591
124. Sauer K, Huang YH, Lin H, Sandberg M, Mayr GW (2009) Phosphoinositide and inositol phosphate analysis in lymphocyte activation. Curr Protoc Immunol. Chapter 11: Unit11.11
125. Adelt S, Plettenburg O, Stricker R, Reiser G, Altenbach H-J, Vogel G (1999) Enzyme-assisted total synthesis of the optical antipodes D-*myo*-inositol 3,4,5-trisphosphate and D-*myo*-inositol 1,5,6-trisphosphate: aspects of their structure-activity relationship to biologically active inositol phosphates. J Med Chem 42:1262–1273
126. Smith RE, MacQuarrie RA, Jope RS (1991) Ion chromatographic determination of inositol tris- and tetrakisphosphates in rat brain. J Chromatogr Sci 29(12):528–531

127. Wilson MS, Bulley SJ, Pisani F, Irvine RF, Saiardi A (2015) A novel method for the purification of inositol phosphates from biological samples reveals that no phytate is present in human plasma or urine. Open Biol 5(3)
128. Pesesse X, Choi K, Zhang T, Shears SB (2004) Signalling by higher inositolpolyphosphates: synthesis of bis-diphosphoinositol tetrakisphosphate ("Insp8") is selectively activated by hyperosmotic stress. JBC 279:43378–43381
129. Choi K, Mollapour E, Choi JH, Shears SB (2008) Cellular energetic status supervises the synthesis of bis-diphosphoinositol tetrakisphosphate independently of amp-activated protein kinase. Mol Pharmacol 74:527–536
130. Gu C, Nguyen HN, Hofer A, Jessen HJ, Dai X, Wang H, Shears SB (2017) The significance of the bifunctional kinase/phosphatase activities of Ppip5ks for coupling inositol pyrophosphate cell-signaling to cellular phosphate homeostasis. J Biol Chem 292:4544–4555
131. Oura T, Murata K, Morita T, Nezu A, Arisawa M, Shuto S, Tanimura A (2016) Highly sensitive measurement of inositol 1,4,5-trisphosphate by using a new fluorescent ligand and ligand binding domain combination. Chembiochem 17(16):1509–1512
132. Sakaguchi R, Tainaka K, Shimada N, Nakano S, Inoue M, Kiyonaka S, Mori Y, Morii T (2010) An in vivo fluorescent sensor reveals intracellular ins(1,3,4,5)P4 dynamics in single cells. Angew Chem Int Ed Engl 49 (12):2150–2153
133. Letcher AJ, Schell MJ, Irvine RF (2008) Do mammals make all their own inositol hexakisphosphate? Biochem J 416(2):263–270
134. Grases F, Costa-Bauza A, Prieto RM (2005) Intracellular and extracellular myo-inositol hexakisphosphate (Insp6), from rats to humans. Anticancer Res 25(3c):2593–2597
135. Perello J, Grases F (2014) Phytate levels in biological fluids of mammals. J Chromatogr B Analyt Technol Biomed Life Sci 960:255–257
136. Irvine RF, Bulley SJ, Wilson MS, Saiardi A (2015) There is no 'conundrum' of Insp6. Open Biol 5(11)
137. Grases F, Isern B, Sanchis P, Perello J, Torres JJ, Costa-Bauza A (2007) Phytate acts as an inhibitor in formation of renal calculi. Front Biosci 12:2580–2587
138. Schultz C, Burmester A, Stadler C (1996) Synthesis, separation, and identification of different inositol phosphates. Subcell Biochem 26:371–413
139. Capolicchio S, Thakor DT, Linden A, Jessen HJ (2013) Synthesis of unsymmetric diphospho-inositol polyphosphates. Angew Chem Int Ed Engl 52:6912–5916
140. Pavlovic I, Thakor DT, Jessen HJ (2016) Synthesis of 2-diphospho-myo-inositol 1,3,4,5,6-pentakisphosphate and a photocaged analogue. Org Biomol Chem 14 (24):5559–5562
141. Wu M, Dul BE, Trevisan AJ, Fiedler D (2013) Synthesis and characterization of non-hydrolysable diphosphoinositol polyphosphate second messengers. Chem Sci 4 (1):405–410
142. Riley AM, Wang H, Weaver JD, Shears SB, Potter BVL (2012) First synthetic analogues of diphosphoinositol polyphosphates: interaction with Ppip5 kinase. Chem Commun 48:11292–11294
143. Wu M, Chong LS, Capolicchio S, Jessen HJ, Resnick AC, Fiedler D (2014) Elucidating diphosphoinositol polyphosphate function with nonhydrolyzable analogues. Angew Chem Int Ed Engl 53:9508–9511
144. Capolicchio S, Wang H, Thakor DT, Shears SB, Jessen HJ (2014) Synthesis of densely phosphorylated Bis-1,5-diphospho-myo-inositol tetrakisphosphate and its enantiomer by bidirectional P-anhydride formation. Angew Chem Int Ed Engl 53:9508–9511
145. Weaver JD, Wang H, Shears SB (2013) The kinetic properties of a human Ppip5k reveal that its kinase activities are protected against the consequences of a deteriorating cellular bioenergetic environment. Biosci Rep 33:228–241
146. Irvine RF, Letcher AJ, Heslop JP, Berridge MJ (1986) The inositol tris/tetrakisphosphate pathway - demonstration of ins(1,4,5) P3 3-kinase activity in animal tissues. Nature 320:631–634
147. Ferris CD, Huganir RL, Supattapone S, Snyder SH (1989) Purified inositol 1,4,5-trisphosphate receptor mediates calcium influx in reconstituted lipid vesicles. Nature 342:87–89
148. Furuichi T, Yoshikawa S, Miyawaki A, Wada K, Maeda N, Mikoshiba K (1989) Primary structure and functional expression of the inositol 1,4,5-trisphosphate-binding protein P_{400}. Nature 342:32–38
149. Rossi AM, Riley AM, Tovey SC, Rahman T, Dellis O, Taylor EJ, Veresov VG, Potter BV, Taylor CW (2009) Synthetic partial agonists

reveal key steps in Ip3 receptor activation. Nat Chem Biol 5(9):631–639

150. McCarren M, Potter BVL, Miller RJ (1989) A metabolically stable analog of 1,4,5-inositol trisphosphate activates a novel K^+ conductance in pyramidal cells of the rat hippocampal slice. Neuron 3:461–471

151. Cullen PJ, Dawson AP, Irvine RF (1995) Purification and characterization of an ins (1,3,4,5)P4 binding protein from pig platelets: possible identification of a novel non-neuronal ins(1,3,4,5)P4 receptor. Biochem J 305:139–143

152. Theibert AB, Estevez VA, Mourney RJ, Marecek JF, Barrow RK, Prestwich GD, Snyder SH (1992) Photoaffinity labeling and characterization of isolated inositol 1,3,4,5-tetrakisphosphate and inositol hexakisphosphate binding proteins. J Biol Chem 267:9071–9079

153. Estevez VA, Prestwich GD (1991) Affinity probes for ins(1,3,4,5)P4 receptors. Tetrahedron 32:1623–1626

154. Cullen PJ, Hsuan JJ, Truong O, Letcher AJ, Jackson TR, Dawson AP, Irvine RF (1995) Identification of a specific ins(1,3,4,5)P4-binding protein as a member of the Gapl family. Nature 376:527–530

155. Battram AM, Durrant TN, Agbani EO, Heesom KJ, Paul DS, Piatt R, Poole AW, Cullen PJ, Bergmeier W, Moore SF, Hers I (2017) The phosphatidylinositol 3,4,5-trisphosphate (pi(3,4,5)P3) binder Rasa3 regulates phosphoinositide 3-kinase (Pi3k)-dependent integrin $\varsigma_{\text{iib}}\beta 3$ outside-in signaling. J Biol Chem 292(5):1691–1704

156. Schurmans S, Polizzi S, Scoumanne A, Sayyed S, Molina-Ortiz P (2015) The Ras/-rap Gtpase activating protein Rasa3: from gene structure to in vivo functions. Adv Biol Regul 57:153–161

157. Voglmaier SM, Keen JH, Murphy J-E, Ferris CD, Prestwich GD, Snyder SH, Theibert AB (1992) Inositol hexakisphosphate receptor identified as the clathrin assembly protein Ap-2. Biochem Biophys Res Commun 187:158–163

158. Krauss M, Kinuta M, Wenk MR, De CP, Takei K, Haucke V (2003) Arf6 stimulates clathrin/Ap 2 recruitment to synaptic membranes by activating phosphatidylinositol phosphate kinase type Igamma. J Cell Biol 162(1):113–124

159. Lemmon MA, Ferguson KM, Abrams CS (2002) Pleckstrin homology domains and the cytoskeleton. FEBS Lett 513(1):71–76

160. Shears SB (2001) Assessing the omnipotence of inositol hexakisphosphate. Cell Signal 13:151–158

161. Torres J, Domínguez S, Cerdá FM, Obal G, Mederos A, Irvine RF, Dìaz A, Kremer C (2005) Solution behaviour of myo-inositol hexakisphosphate in the presence of multivalent cations. Prediction of a neutral pentamagnesium species under cytosolic/nuclear conditions. J Inorg Biochem 99:828–840

162. Ali N, Craxton A, Shears SB (1993) Hepatic ins(1,3,4,5)P_4 3-phosphatase is compartmentalized inside endoplasmic reticulum. J Biol Chem 268:6161–6167

163. Van Der KJ, Van Haastert PJ (1995) Desalting inositolpolyphosphates by dialysis. Anal Biochem 225(1):183–185

164. Abdullah M, Hughes PJ, Craxton A, Gigg R, Desai T, Marecek JF, Prestwich GD, Shears SB (1992) Purification and characterization of inositol 1,3,4-trisphosphate 5/6-kinase from rat liver using an inositol hexakisphosphate affinity column. J Biol Chem 267:22340–22345

165. Jiao C, Summerlin M, Bruzik KS, Hanakahi L (2015) Synthesis of biotinylated inositol hexakisphosphate to study DNA double-strand break repair and affinity capture of Ip6-binding proteins. Biochemistry 54 (41):6312–6322

166. Wu M, Chong LS, Perlman DH, Resnick AC, Fiedler D (2016) Inositol polyphosphates intersect with signaling and metabolic networks via two distinct mechanisms. Proc Natl Acad Sci U S A 113:E6757–E6765

167. Vajanaphanich M, Schultz C, Rudolf MT, Wasserman M, Enyedi P, Craxton A, Shears SB, Tsien RY, Barrett KE, Traynor-Kaplan AE (1994) Long-term uncoupling of chloride secretion from intracellular calcium levels by ins(3,4,5,6)P_4. Nature 371:711–714

168. Li W, Schultz C, Llopis J, Tsien RY (1997) Membrane-permeant esters of inositol polyphosphates, chemical synthesis and biological applications. Tetrahedron 53:12017–12040

169. Li W, Llopis J, Whitney M, Zlokarnik G, Tsien RY (1998) Cell-Permeant caged Insp3 Ester shows that Ca2+ spike frequency can optimize gene expression. Nature 392 (6679):936–941

170. Pavlovic I, Thakor DT, Vargas JR, McKinlay CJ, Hauke S, Anstaett P, Camuna RC, Bigler L, Gasser G, Schultz C, Wender PA, Jessen HJ (2016) Cellular delivery and photochemical release of a caged inositol-pyrophosphate induces Ph-domain translocation in Cellulo. Nat Commun 7:10622

171. Chakraborty A, Koldobskiy MA, Bello NT, Maxwell M, Potter JJ, Juluri KR, Maag D, Kim S, Huang AS, Dailey MJ, Saleh M, Snowman AM, Moran TH, Mezey E, Snyder SH (2010) Inositol pyrophosphates inhibit Akt signaling, thereby regulating insulin sensitivity and weight gain. Cell 143(6):897–910

172. Bhandari R, Saiardi A, Ahmadibeni Y, Snowman AM, Resnick AC, Kristiansen TZ, Molina H, Pandey A, Werner JK Jr, Juluri KR, Xu Y, Prestwich GD, Parang K, Snyder SH (2007) Protein pyrophosphorylation by inositol pyrophosphates is a posttranslational event. Proc Natl Acad Sci U S A 104:15305–15310

173. Saiardi A, Bhandari A, Resnick R, Cain A, Snowman AM, Snyder SH (2004) Inositol pyrophosphate: physiologic phosphorylation of proteins. Science 306:2101–2105

174. Williams FJ, Fiedler D (2015) A fluorescent sensor and gel stain for detection of pyrophosphorylated proteins. ACS Chem Biol 10 (9):1958–1963

175. Poyner DR, Cooke F, Hanley MR, Reynolds DJM, Hawkins PT (1993) Characterization of metal ion-induced 3Hinositol hexakisphosphate binding to rat cerebellar membarnes. J Biol Chem 268:1032–1038

176. Macbeth MR, Schubert HL, Vandemark AP, Lingam AT, Hill CP, Bass BL (2005) Inositol hexakisphosphate is bound in the Adar2 core and required for Rna editing. Science 309 (5740):1534–1539

177. Sakaguchi R, Endoh T, Yamamoto S, Tainaka K, Sugimoto K, Fujieda N, Kiyonaka S, Mori Y, Morii T (2009) A single circularly permuted Gfp sensor for Inositol-1,3,4,5-tetrakisphosphate based on a split Ph domain. Bioorg Med Chem 17 (20):7381–7386. https://doi.org/10.1016/j.bmc.2009.08.015

178. Worley J, Luo X, Capaldi AP (2013) Inositol pyrophosphates regulate cell growth and the environmental stress response by activating the Hdac Rpd3l. Cell Rep 3:1476–1482

179. Brehm MA, Wundenberg T, Williams J, Mayr GW, Shears SB (2013) A non-catalytic role for inositol 1,3,4,5,6-pentakisphosphate 2-kinase in the synthesis of ribosomal Rna. J Cell Sci 126:437–444

180. Kim S, Kim SF, Maag D, Maxwell MJ, Resnick AC, Juluri KR, Chakraborty A, Koldobskiy MA, Cha SH, Barrow R, Snowman AM, Snyder SH (2011) Amino acid signaling to Mtor mediated by inositol polyphosphate multikinase. Cell Metab 13(2):215–221

181. Rao F, Xu J, Khan AB, Gadalla MM, Cha JY, Xu R, Tyagi R, Dang Y, Chakraborty A, Snyder SH (2014) Inositol hexakisphosphate kinase-1 mediates assembly/disassembly of the Crl4-signalosome complex to regulate DNA repair and cell death. Proc Natl Acad Sci U S A 111(45):16005–16010

182. Padmanabhan U, Dollins DE, Fridy PC, York JD, Downes CP (2009) Characterization of a selective inhibitor of inositol hexakisphosphate kinases: use in defining biological roles and metabolic relationships of inositol pyrophosphates. J Biol Chem 284:10571–10582

183. Ghoshal S, Zhu Q, Asteian A, Lin H, Xu H, Ernst G, Barrow JC, Xu B, Cameron MD, Kamenecka TM, Chakraborty A (2016) Tnp [N2-(M-Trifluorobenzyl), N6-(P-Nitrobenzyl)purine] ameliorates diet induced obesity and insulin resistance *Via* inhibition of the Ip6k1 pathway. Molecular Metabolism 5:903–917

Part I

Isolation and Measurement

Chapter 2

Quantitation of Inositol Phosphates by HPLC-ESI-MS

Quynh H. Duong and Ronald B. Pegg

Abstract

The coupling of anion exchange high-pressure liquid chromatography (HPLC) with electrospray ionization mass spectrometry (ESI-MS) allows for the simultaneous detection of the six forms of inositol phosphate (InsP). Here we describe a rapid quantitative analysis of InsPs by HPLC-ESI-MS, which can be applied to a wide array of sample types. With this method, InsPs could be separated and detected within 20 min of sample injection. The detection limit was as low as 25 pmol (i.e., ca. 2 nmol/g sample) for each type of InsP, which is particularly important for analytes that are often present at low abundance in nature.

Key words Inositol phosphate, Phytic acid, HPLC, Ion-exchange HPLC, HPLC-ESI-MS

1 Introduction

Inositol phosphate (InsP) refers to a family of compounds where a *myo*-inositol molecule is phosphorylated at one or more locations on the six-carbon ring. The degree of phosphorylation can range anywhere from one to six, resulting in *myo*-inositol mono-, bis-, tris-, tetrakis-, pentakis, or hexakisphosphate ($InsP_{1–6}$). The phosphate group/s can lose protons to give the compound an anionic property and allow them to be separated on an anion-exchange chromatographic column. While many existing methods have been developed to separate InsPs by high-pressure liquid chromatography (HPLC), the detection of these compounds for quantitative purpose continues to be a challenge. Because InsPs lack a characteristic spectrophotometric absorbance maximum, post-column detection usually requires a cumbersome off-line approach to generate colored/fluorescent complexes [1–4] or to cause the disappearance of color in a colored reagent [5, 6].

In order to achieve on-line detection and greater time efficiency, recent methods have explored the employment of mass spectrometry (MS) for the analysis of InsPs. The coupling of chromatography with MS or tandem MS has been shown to be applicable to a wide array of samples, ranging from environmental,

Gregory J. Miller (ed.), *Inositol Phosphates: Methods and Protocols*, Methods in Molecular Biology, vol. 2091,
https://doi.org/10.1007/978-1-0716-0167-9_2,

biological, to food materials [7–10]. Here, we describe a method utilizing HPLC and MS to rapidly separate and detect $InsP_{1–6}$ without differentiating their stereoisomers. With this method, we have successfully measured the content of $InsP_{1–6}$ in several complex plant materials, including eight types of tree nuts (cashews, Brazil nuts, macadamias, hazelnuts, walnuts, pecans, pistachios, almonds) and three types of grain components, typically rich in InsPs (corn germ, rice bran, wheat cotyledon) [11, 12]. The detection limit for each InsP was 25 pmol (ca. 2 nmol/g sample), highlighting the possibility to apply this method for many types of samples that require high sensitivity.

2 Materials

Prepare all solutions in HPLC-grade water and at room temperature.

1. Methanol–water solution: 5% HPLC-grade methanol in HPLC water (*v/v*). Store at room temperature. This also serves as mobile phase A.
2. Mobile phase B: 200 mM ammonium carbonate. Dissolve 19.218 g $(NH_4)_2CO_3$ (≥99% purity) in 1 L of water. Store at room temperature.
3. Extracting solvent: 3.2 M acetic acid. In a 100 mL volumetric flask, add 50 mL of water, then add 18.4 mL of glacial acetic acid. Fill the flask to mark with water. Store at room temperature.
4. Adenosine 5′-monophosphate (AMP) standard: 250 μM AMP. Dissolve 8.7 mg of AMP in 100 mL of HPLC water. Store at −20 °C or below.
5. Analytical standards: analytical standards of $InsP_{1–6}$ should have ≥98% purity. The method does not differentiate stereoisomers. Dilute each standard with HPLC water and combine with the AMP standard to make a series of standard solutions: the recommended concentrations for each InsP are 5, 20, 30, 40, 50, 60, 80, and 100 μM (*see* **Note 1**); the final concentration of AMP in each solution is 25 μM. Store these solutions at −20 °C or below.
6. Na_2EDTA-NaOH solution: 0.11 M Na_2EDTA and 0.75 M NaOH. Dissolve 10.23 g Na_2EDTA with 7.5 g NaOH in 250 mL of HPLC water. Store at room temperature.

3 Methods

3.1 HPLC-ESI-MS System

1. Equip the HPLC with a BioBasic anion-exchange column (2.1 × 150 mm, 5-μm particle size, Thermo Fisher Scientific, Waltham, MA, USA) or equivalent (*see* **Note 2**).
2. Install mobile phases A and B into the system. Set up the HPLC program to include a gradient starting with 4 min of mobile phase A, followed by an increase of B from 0 to 20% over 2 min, and lastly an increase of B from 20 to 55% over 14 min. It is suggested that a 20-min regeneration period with 100% mobile phase A be included at the end of each run. Set the flow rate throughout the program at 200 μL/min. The analysis can be performed at room temperature.
3. The samples will be detected with a quadrupole time-of-flight mass spectrometer (MS) with an electrospray ionization (ESI) source (*see* **Note 3**). Set the ESI to the negative mode. To optimize the settings of the system, a mixture of InsP and AMP standards should be injected directly into the MS analyzer and a manual tune or auto-tune should be performed (*see* **Note 4**). In the negative mode, the mass-to-charge ratio (*m/z*) corresponding to AMP is 346 and to $InsP_{1–6}$ are 259, 339, 419, 499, 579, and 659, respectively.
4. Collect data with a mass analysis software (*see* **Note 5**). Representative elution peaks of AMP and $InsP_{1–6}$ are depicted in Fig. 1.

3.2 Construction of Standard Curves

1. Analyze the InsP standards at different concentrations with the HPLC-ESI-MS system. The recommended injection volume is 5 μL.
2. Using the elution time and *m/z* information, identify the elution peak of AMP and of each InsP in these runs. Calculate the area of these peaks.
3. Calculate the ratio of each InsP's peak area over AMP's peak area. Plot this ratio against the quantity of InsP in the injection volume (pmol). An example of the standard curves of $InsP_{1–6}$ can be found in Fig. 2.

3.3 Sample Analysis

Sample preparation: Solid samples should be ground into a fine powder. Samples with a high fat content should first be defatted with an appropriate method, and the fat content needs to be recorded so that it can be taken into account in the calculation of InsP concentrations (*see* **Note 6**).

1. Mix 500 mg of sample with 3 mL of 3.2 M acetic acid.
2. Blend at 15,000 rpm for 60 s with a homogenizer.

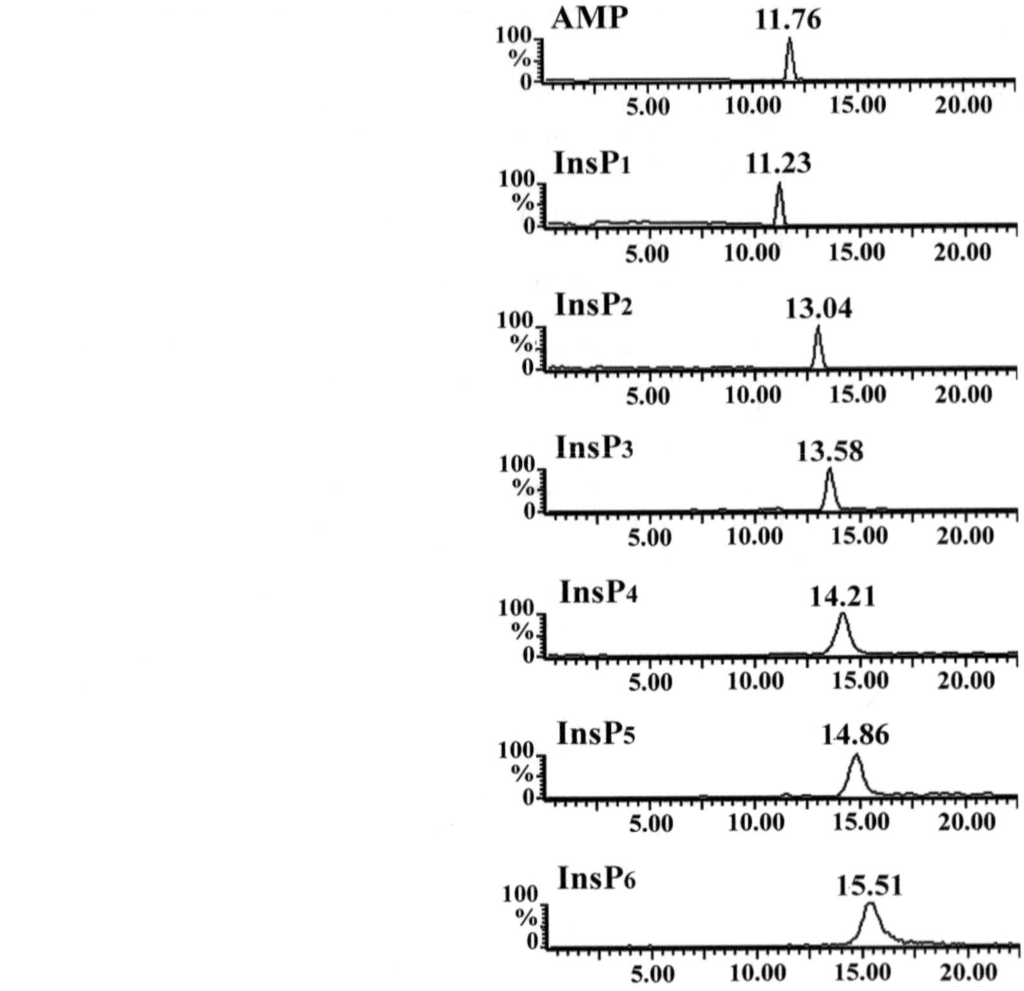

Fig. 1 Representative elution peaks of $InsP_{1-6}$ and adenosine 5′-monophosphate detected in commercial standards

3. Shake the sample for 3 h, then centrifuge until the supernatant is clear (*see* **Note 7**).
4. Combine half a milliliter of the supernatant with 1 mL of the Na_2EDTA—NaOH solution in a container fit for a SpeedVac System. Evaporate the sample to dryness in the SpeedVac (*see* **Note 8**).
5. Reconstitute the residue in 500 μL methanol–water and sonicate to facilitate dissolution.
6. Filter the sample through a 0.22-μm Phenex™-RC nylon syringe filter or equivalent.
7. Combine the sample with the AMP standard solution. The final concentration of AMP should be 25 μM.

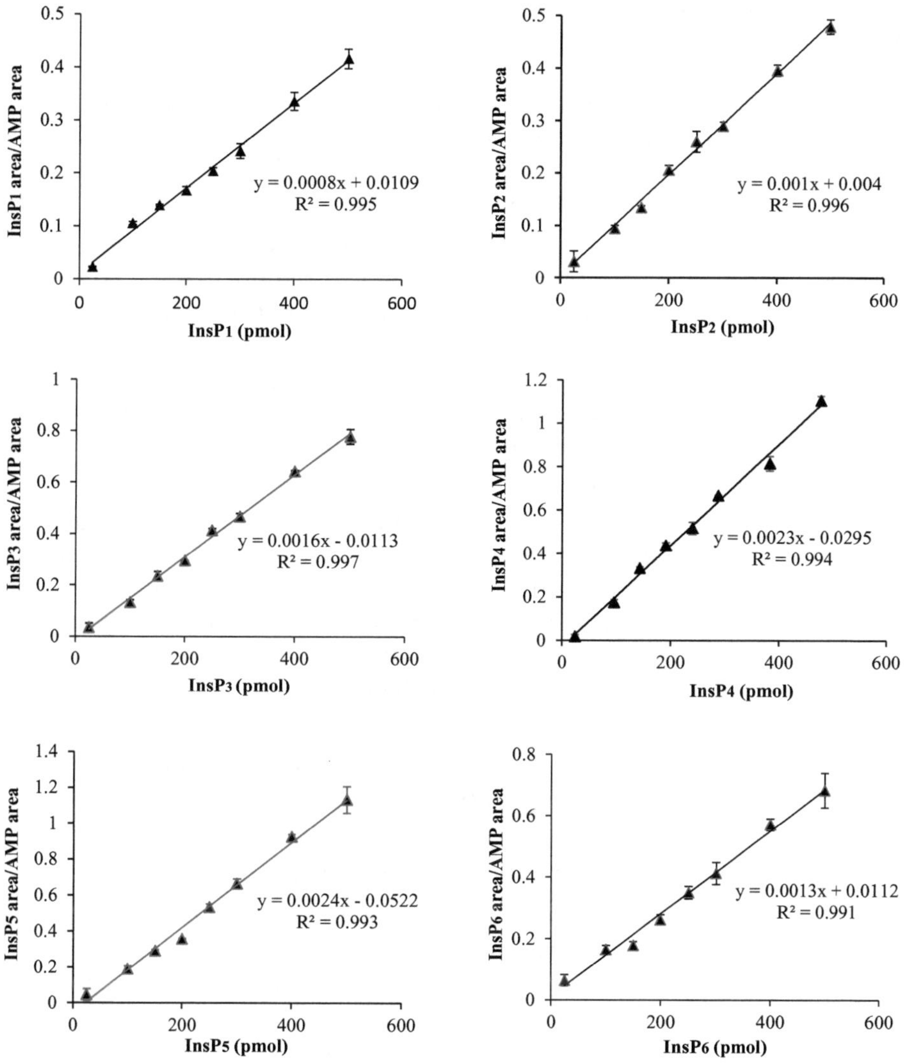

Fig. 2 Example calibration curves of $InsP_{1-6}$

8. Analyze the sample with the HPLC/ESI/MS system. The recommended injection volume is 5 μL.
9. Using the elution time and *m/z* information, identify the elution peak of AMP and of each InsP in these runs. Calculate the area of these peaks.
10. Calculate the ratio of each InsP peak area over AMP's peak area. Use the standard curve equations to calculate the amount of each InsP in the injected volume and in the original sample.

4 Notes

1. More concentrations can be added as necessary. For example, plant seeds often contain very high amounts of $InsP_6$ compared to other InsPs, in which case higher concentrations of $InsP_6$ can be added.
2. Using a different column can significantly alter the retention time of the analytes and may require that the HPLC gradient be adjusted to optimize the separation. This also applies to columns that belong to the same series that we described, but have different dimensions—such as the BioBasic anion-exchange column (0.5 × 150 mm, 5 mm), which Liu and others equipped on an HPLC system with a capillary pump in a similar study to this one [10].
3. In our study, we employed a Waters Micromass quadrupole time-of-flight mass spectrometer. In a study of similar nature, other authors have used different mass analyzers, including a TSQ Quantum Ultra AM triple quadrupole mass spectrometer and a Discovery Max triple quadrupole mass spectrometer (Thermo Scientific, San Jose, CA, USA) [10].
4. An example of the settings is as follows: spray voltage at 2500 V, ion-transfer capillary temperature at 300 °C, and nitrogen gas flow at 450 L/h.
5. For instance, we used the MassLynx software (Version 4.1, Waters Corporation, Milford, MA, USA).
6. For samples with a very high fat content such as tree nuts, we found that flash-freezing the sample in liquid nitrogen before grinding it with a strong grinder (such as a coffee grinder) in short pulses (roughly 20 s each) was most effective for reducing the particle size of the sample. Solid samples can be defatted with hexanes in a Soxhlet apparatus. Afterward, residual hexanes in the ground, defatted sample can be removed by placing the sample in a fume hood for a period of time.
7. An example of the centrifugation program is 3345 × *g* for 20 min. In a similar study, Liu and others used a program with settings of 5000 × *g*, 20 min, and 4 °C. In our experience, the effect of temperature setting on the efficiency of this step varies depending on the sample. With the tree nut and grain samples we worked, room temperature was more effective than low temperatures such as 4 °C.
8. For samples that tend to get compressed in the SpeedVac and cause problems in the reconstitution step, this drying step can be carried out instead using a freeze-dryer or other low-heat drying system.

Acknowledgments

Financial supports for this research by The Graduate School of UGA and the Almond Board of California are gratefully acknowledged. Use of the Proteomics and Mass Spectrometry (PAMS) core facility with the assistance of Dr. Kevin D. Clark in UGA's Department of Chemistry, Athens, GA, is greatly appreciated.

References

1. Cao S, Dong N, Chen J (2011) Synchronous fluorescence determination of phytic acid in foodstuffs and urine based on replacement reaction. Phytochem Anal 22:119–123
2. Chen Y, Chen J, Ma K, Cao S, Chen X (2007) Fluorimetric determination of phytic acid in urine based on replacement reaction. Anal Chim Acta 605:185–191
3. Irth H, Lamoree M, de Jong GJ, Brinkman UATh, Frei RW, Kornfeldt RA et al (1990) Determination of D-*myo*-1,2,6-inositol trisphosphate by ion-pair reversed-phase liquid chromatography with post-column ligand exchange and fluorescence detection. J Chromatogr A 499:617–625
4. March JG, Simonet BM, Grases F (1999) Fluorimetric determination of phytic acid based on the activation of the oxidation of 2,2′-dipyridyl ketone hydrazone catalysed by Cu(II). Analyst 124:897–900
5. Rounds MA, Nielsen SS (1993) Anion-exchange high-performance liquid chromatography with post-column detection for the analysis of phytic acid and other inositol phosphates. J Chromatogr A 653:148–152
6. Guse AH, Greiner E, Emmrich F, Brand K (1993) Mass changes of inositol 1,3,4,5,6-pentakisphosphate and inositol hexakisphosphate during cell cycle progression in rat thymocytes. J Biol Chem 268:7129–7133
7. Sjöberg PJR, Thelin P, Rydin E (2016) Separation of inositol phosphate isomers in environmental samples by ion-exchange chromatography coupled with electrospray ionization tandem mass spectrometry. Talanta 161:392–397
8. Guo J, Shi Y, Xu C, Zhong R, Zhang F, Zhang T et al (2016) Quantification of plasma myo-inositol using gas chromatography–mass spectrometry. Clin Chim Acta 460:88–92
9. Zhang S, Yang W, Zhao Q, Zhou X, Fan Y, Chen R (2017) Rapid method for simultaneous determination of inositol phosphates by IPC-ESI–MS/MS and its application in nutrition and genetic research. Chromatographia 80:275–286
10. Liu X, Villalta PW, Sturla SJ (2009) Simultaneous determination of inositol and inositol phosphates in complex biological matrices: quantitative ion-exchange chromatography/tandem mass spectrometry. Rapid Commun Mass Spectrom 23:705–712
11. Duong QH, Clark KD, Lapsley KG, Pegg RB (2017) Determination of *myo*-inositol phosphates in tree nuts and grain fractions by HPLC-ESI-MS. J Food Compos Anal 59:74–82
12. Duong QH, Clark KD, Lapsley KG, Pegg RB (2017) Quantification of inositol phosphates in almond meal and almond brown skins by HPLC/ESI/MS. Food Chem 229:84–92

Acknowledgments

Chapter 3

Isolation of Inositol Hexakisphosphate from Soils by Alkaline Extraction and Hypobromite Oxidation

Benjamin L. Turner

Abstract

Inositol hexakisphosphates are extracted from soil in strong alkali and isolated from other organic phosphates by hypobromite oxidation. The procedure yields the four stereoisomeric forms of inositol hexakisphosphate in a form suitable for spectroscopic or chromatographic identification.

Key words Inositol phosphate, NMR spectroscopy, Phytate, Phytic acid, Soil, Sediment, Stereoisomers

1 Introduction

Inositol phosphates can constitute a considerable fraction of the soil organic phosphorus, occurring predominantly in four stereoisomeric forms of inositol hexakisphosphate [1, 2]. The most abundant is *myo*-inositol hexakisphosphate, commonly known as phytic acid (the free acid) or phytate (the salts), which forms the majority of the phosphorus in plant seeds [3]. The *scyllo*-, *neo*, and D-*chiro* isomers (Fig. 1) are typically present in smaller amounts and appear to originate from microorganisms [2]. Inositol hexakisphosphate is stable in soils, but can be extracted in strong alkali in conjunction with a chelating agent such as ethylenediaminetetraacetate [4, 5]. Indeed, a combination of 0.25 M NaOH and 50 mM EDTA is now a routine extraction for soil organic phosphorus [6]. However, NaOH–EDTA extracts contain considerable concentrations of organic matter, which can complicate subsequent detection by chromatographic or spectroscopic procedures. For example, the abundance of compounds in the phosphomonoester region of solution ^{31}P NMR spectra can preclude the identification and quantification of the inositol hexakisphosphate stereoisomers (Fig. 2). To overcome this, inositol hexakisphosphates can be isolated by hypobromite oxidation, which destroys all organic

Gregory J. Miller (ed.), *Inositol Phosphates: Methods and Protocols*, Methods in Molecular Biology, vol. 2091,
https://doi.org/10.1007/978-1-0716-0167-9_3, © Springer Science+Business Media, LLC, part of Springer Nature 2020

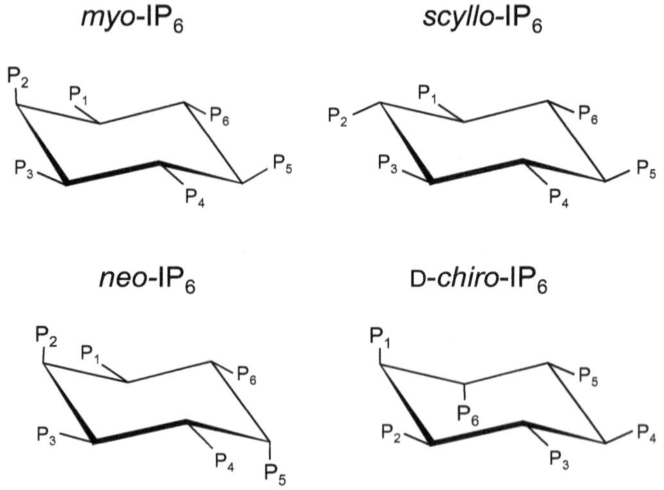

Fig. 1 The four stereoisomers of inositol hexakisphosphate that occur in soils. P_1 to P_6 are phosphate groups, with the subscript numbers indicating the positions on the inositol ring. Conformations are shown at environmental pH

matter and organic phosphates other than the inositol hexakisphosphates and some lower-order inositol esters [4, 5, 7, 8]. The procedure yields the compounds in a form suitable for identification by chromatographic or spectroscopic analysis. Total inositol hexakisphosphate can be determined by conventional wet chemical analysis [9].

2 Materials

Prepare all solutions in deionized water purified to 18 MΩ cm^{-1} at 25 °C. Reagents can be stored at ambient laboratory temperature unless stated otherwise. Ensure that waste disposal complies with local regulations. Clean glassware by washing in phosphate-free detergent or dilute hydrochloric acid.

2.1 Alkaline Extraction

Extraction solution: 0.25 M NaOH and 50 mM EDTA. Dissolve 10 g of NaOH and 18.61 g of Na_2EDTA in 500 mL of deionized water in a 1 L beaker. Dilute to 1 L in a volumetric flask and mix well (*see* **Note 1**).

2.2 Hypobromite Oxidation and Isolation of Inositol Phosphates

Bromine, Br_2 (Aldrich 470864, Reagent plus, 99.99%; Sigma-Aldrich 277576, ACS reagent ≥99.5%).

Amberlite® IR120 cation exchange resin, hydrogen form (*see* **Note 2**).

Sodium hydroxide pellets.

Sodium hydroxide: 10 M solution in water.

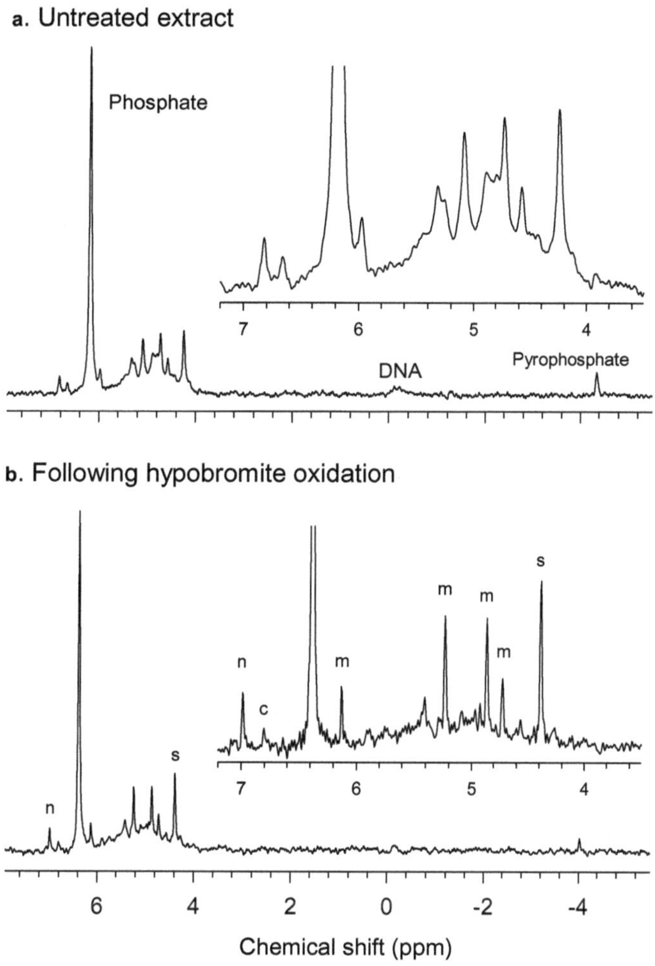

Fig. 2 Example solution ^{31}P NMR spectra of a soil NaOH–EDTA extract with no pretreatment (**a**) and following hypobromite oxidation (**b**). Spectrum **b** shows the marked increase in resolution in the phosphomonoester region (inset spectra) following pretreatment, which improves quantification of inositol hexakisphosphates [5]. The signals are as follows: (**c**), the two equatorial phosphates of D-*chiro*-inositol hexakisphosphate; m, the four signals of *myo*-inositol hexakisphosphate; n, the four equatorial phosphates of *neo*-inositol hexakisphosphate; s, *scyllo*-inositol hexakisphosphate [4]

Hydrochloric acid: 6 M solution in water.

Barium acetate solution: 10% w/v in water.

Ethanol: 50% solution in deionized water.

Potassium dihydrogen phosphate solution: dry for 24 h at 105 °C and allow to cool in a desiccator, then dissolve 4.394 g in deionized water and dilute to 1 L in a volumetric flask.

2.3 Quantification of Inositol Hexakisphosphate

Ammonium molybdate solution: Dissolve 6.0 g ammonium molybdate in 200 mL of deionized water.

Potassium antimony tartrate solution: Dissolve 0.146 g of potassium antimony tartrate in 50 mL of deionized water.

Sulfuric acid solution: 2.5 M in water. Concentrated sulfuric acid is highly corrosive. Wear gloves and goggles during handling and prepare dilute solutions in a fume hood.

Mixed reagent: combine 500 mL of 2.5 M sulfuric acid with the ammonium molybdate and potassium antimony tartrate solutions and dilute to 1 L with deionized water in a volumetric flask. Store in a dark bottle in a cool dark place.

Working reagent: Dissolve 0.528 g of ascorbic acid in 100 mL of mixed reagent. The solution must be made fresh each working day as it is stable for <24 h.

Phenolphthalein indicator: 1% solution in ethanol. Dissolve 1.0 g phenolphthalein in 60 mL ethanol in a 100 mL volumetric flask. Dilute to the mark with deionized water and mix well.

Sulfuric acid: 3.0 M solution in water.

Sodium hydroxide: 1.0 M solution in water.

Nitric acid: 2% solution in water.

3 Methods

3.1 Extraction of Inositol Phosphates from Soil

1. Weigh 1.50 ± 0.01 g of soil (dried at ≤ 40 °C and ground to a fine powder) into a 50 mL centrifuge tube. Add 30 mL of NaOH–EDTA solution, cap the tube, and shake by hand to disperse the soil (*see* **Note 3**).
2. Cap the tube and incubate in a shaking water bath at 85 °C for 2 h (*see* **Note 4**).
3. Allow the samples to cool to room temperature, then centrifuge at 8000 × *g* for 30 min and decant the supernatant (*see* **Note 5**).

3.2 Hypobromite Oxidation

1. Pipet 10 mL of the NaOH–EDTA extract into a 100 mL glass beaker and add 2.0 g solid NaOH to bring the pH of the solution to >13. Very dark extracts (i.e., with high organic matter) may require more NaOH during the following steps if the pH drops below 13. This can be added in 1 mL aliquots of 10 M NaOH. Include a blank containing only NaOH–EDTA extraction solution.
2. Cool the bottle of bromine in the fridge and then place in an ice-bath under a fume hood (*see* **Note 6**). Add 0.5 mL pure bromine to each extract using a disposable dripping pipette.

The extracts should begin to effervesce. Very dark extracts may require an additional 0.5 mL aliquot of bromine. Allow the reaction to proceed for 1 h at room temperature.

3. Cover the beaker with a watch glass and place on a hot plate. Heat the sample until it boils to complete the reaction (the temperature of the solution will be approx. 140 °C). Boil the samples for 5 min and then remove from the heat and allow to cool. The sample should be clear/yellow.
4. Add 6 M HCl in 1 mL aliquots to acidify the sample to <pH 3. This should require at least 6 mL of HCl, but more if additional NaOH has been added. The sample should change from yellow to orange and excess bromine gas should be released (*see* **Note 7**).
5. Return the sample to the hot plate and boil for 5 min until no further bromine gas is released, then allow to cool. The sample should be yellow.
6. Add 10 M NaOH drop wise to adjust to approximately pH 8.5. The color should change from yellow to clear, although some samples will remain yellow.

3.3 Precipitation of Inositol Phosphates

1. Adjust the volume of the sample to 20 mL with water, and then add 10 mL of 50% ethanol and 10 mL of 10% barium acetate. Heat for 10 min at 50–60 °C and then allow to cool.
2. Transfer the contents of the beaker to a 250 mL centrifuge bottle and centrifuge to isolate the precipitated phosphates (1500 × *g*, 10 min) (*see* **Note 8**). Discard the supernatant.
3. Wash the precipitate with 20 mL of 50% ethanol, shaking by hand to dislodge the precipitate from the side of the bottle. Centrifuge as above and discard the supernatant.
4. Suspend the precipitate in 20 mL deionized water with approximately 20 mL of Amberlite ion exchange resin and shake for 10 min. If the solution is still cloudy or a white precipitate is visible, add additional resin and shake again.
5. When the solution is clear, filter through a 0.2 μm syringe filter (e.g., Whatman cellulose-acetate), washing the resin with a few mL of deionized water. Neutralize the filtrate by dropwise addition of 1 M NaOH and dilute to 25 mL in a volumetric flask.

3.4 Quantification of Total Inositol Hexakisphosphate

1. Measure inorganic phosphate: Pipet a 1 mL aliquot of the solution containing inositol hexakisphosphates in a 50 mL volumetric flask and neutralize (*see* **Note 9**). Add 8 mL of molybdate working reagent and dilute to the mark with deionized water, inverting several times to mix. Read the absorbance after at least 12 min at 880 nm against calibration standards between

0 and 25 mg P L^{-1} diluted in the same manner as the samples (*see* **Note 10**).

2. Measure total phosphorus: Prepare a separate sample for total phosphorus determination by diluting a 1 mL aliquot with 9 mL of 2% HNO_3 and analyzing by inductively coupled plasma optical-emission spectrometry (ICP–OES) against calibration standards between 0 and 10 mg P L^{-1} prepared in 2% HNO_3 (*see* **Note 11**).
3. Calculate total inositol hexakisphosphate as the difference between total phosphorus determined by ICP–OES and inorganic phosphate determined by molybdate colorimetry, correcting for the dilution factors, the total volume of solution containing the inositol phosphates, and the oven-dry weight of soil in the original extraction.

3.5 Preparation of Inositol Phosphates for Detection by Chromatography or NMR Spectroscopy

1. The remaining sample can be preconcentrated for analysis by chromatography or spectroscopy by gentle heating, rotary evaporation, or freezing, lyophilization, and redissolving in a solution containing 1.0 M NaOH + 0.1 M Na_2EDTA.

4 Notes

1. Use disodium salt form of EDTA rather than the free acid, which changes the pH and the sodium concentration of the solution and therefore influences the efficiency of soil organic phosphorus extraction.
2. Amberlite resin should be prepared initially and regenerated after use by washing sequentially in water (twice), 1 M HCl (twice), and again in water (three times), with each wash using five times the volume of resin.
3. A 1:50 soil to solution ratio was used in the original development of the NaOH–EDTA extraction procedure [10]. Subsequently, extraction has been performed at a narrower ratio (1:10 or 1:20) to increase phosphorus concentrations for NMR spectroscopy [6, 11]. Whether this influences the extraction of inositol hexakisphosphate is not well understood.
4. An alternative is to incubate the samples in an oven at 85 °C with occasional swirling by hand. Extraction at 85 °C for 2 h was used in the original development of the NaOH–EDTA extraction procedure [10]. Subsequently, extraction has been performed at ambient laboratory temperature on a shaking machine [6, 12]. Whether this influences the extraction of inositol hexakisphosphate is not well understood.

5. An alternative to centrifuging is to vacuum filter the solution through a glass fiber filter paper (e.g., Whatman GF/B). To avoid dislodging soil after centrifuging, remove the solution in 5 mL aliquots using a pipette, leaving the final 5 mL of solution in the centrifuge tube.
6. Bromine is corrosive and toxic, so handle with extreme caution. Placing the bottle of bromine in an ice bath in the fume hood prevents gaseous loss and facilitates transfer to the samples.
7. If no gas is released then add additional HCl until the pH reaches <3. If necessary, measure the pH during the acidification to confirm it is <3.
8. Alternatively the samples can be left to stand overnight at ambient laboratory temperature and centrifuged in the morning.
9. The solution should be at approximately neutral pH. If not, neutralize the solution by adding a drop of phenolphthalein solution—if pink, add 3 M H_2SO_4 dropwise until the solution turns clear; if clear, add 1.0 M NaOH dropwise until pink, then add one drop of 3 M H_2SO_4. The solution should be clear.
10. The volume of extract will depend on the phosphorus concentration, but should not exceed 25 μg of phosphate.
11. If ICP–OES is not available, an alternative is to digest samples by acid digestion or persulfate oxidation with phosphate detection by molybdate colorimetry as described above [13].

References

1. Cosgrove DJ (1980) Inositol phosphates - their chemistry, biochemistry and physiology. Elsevier Scientific, Amsterdam
2. Turner BL (2007) Inositol phosphates in soil: amounts, forms and significance of the phosphorlyated inositol stereoisomers. In: Turner BL, Richardson AE, Mullaney EJ (eds) Inositol phosphates: linking agriculture and the environment. CAB International, Wallingford, pp 186–207
3. Raboy V (2003) *myo*-Inositol-1,2,3,4,5,6-hexa*kis*phosphate. Phytochemistry 64:1033–1043
4. Turner BL, Cheesman AW, Godage HY, Riley AM, Potter BVL (2012) Determination of *neo*- and D-*chiro*-inositol hexakisphosphate in soils by solution ^{31}P NMR spectroscopy. Environ Sci Technol 46:4994–5002. https://doi.org/10.1021/es204446z
5. Turner BL, Richardson AE (2004) Identification of *scyllo*-inositol phosphates in soils by solution phosphorus-31 nuclear magnetic resonance spectroscopy. Soil Sci Soc Am J 68 (3):802–808
6. Cade-Menun BJ, Preston CM (1996) A comparison of soil extraction procedures for ^{31}P NMR spectroscopy. Soil Sci 161(11):770–785
7. Wrenshall CL, Dyer WJ (1941) Organic phosphorus in soils: II. The nature of the organic phosphorus compounds. A. Nucleic acid derivatives. B. Phytin. Soil Sci 51:235–248
8. Irving GCJ, Cosgrove DJ (1981) The use of hypobromite oxidation to evaluate two current methods for the estimation of inositol polyphosphates in alkaline extracts of soils. Commun Soil Sci Plant Anal 12(5):495–509

9. Murphy J, Riley JP (1962) A modified single solution method for the determination of phosphate in natural waters. Anal Chim Acta 27:31–36
10. Bowman RA, Moir JO (1993) Basic EDTA as an extractant for soil organic phosphorus. Soil Sci Soc Am J 57:1516–1518
11. Turner BL, Cade-Menun BJ, Condron LM, Newman S (2005) Extraction of soil organic phosphorus. Talanta 66(2):294–306
12. Turner BL, Mahieu N, Condron LM (2003) The phosphorus composition of temperate pasture soils determined by NaOH-EDTA extraction and solution ^{31}P NMR spectroscopy. Org Geochem 34(8):1199–1210
13. Rowland AP, Haygarth PM (1997) Determination of total dissolved phosphorus in soil solutions. J Environ Qual 26:410–415

Chapter 4

Detection of Inositol Phosphates by Split PH Domains

Reiko Sakaguchi, Shunsuke Tajima, Yasuo Mori, and Takashi Morii

Abstract

The pleckstrin homology (PH) domain is a family of structurally conserved proteins which can bind inositol phosphate derivatives. Some proteins involved in cellular signaling and cytoskeletal organization possess split PH domains that assemble into a structure which can bind specific inositol phosphates. Here we describe the design of split PH domain from a structurally well-characterized PH domain of phospholipase C (PLC) δ_1 and Bruton's tyrosine kinase (Btk), which selectively bind Ins(1,4,5)P_3 and Ins(1,3,4,5)P_4, respectively. The PH domains fold into a functional structure when the split halves are brought to close proximity, and can be utilized to detect specific inositol phosphate of interest.

Key words Inositol phosphates, PH domain, Structure-based design, Fluorescence

1 Introduction

The pleckstrin homology (PH) domain is a family of structurally conserved small protein domain of approximately 13–15 kDa (~120 amino acids) [1]. PH domains are relatively basic so they can bind inositol phosphate derivatives by charge interaction. It is known that some proteins involved in cellular signaling and cytoskeletal organization including PLCγ1 [2, 3], syntrophin [4], Rho Kinase [5], Myosin X [6], ESCRT-II [7], possess split PH domains that assembles into a structure which exerts certain function [8–14]. We have designed an artificially split PH domain from a structurally well-characterized PH domain of phospholipase C (PLC) δ_1 [15] by dissecting the PH domain and tethering a coiled-coil module to each subunit to successfully reassemble the split PH domain that retains the functions of the parent PH domain (Fig. 1) [16]. In this case, PLCδ_1 PH domain was divided into two subunits at the β2/β3 loop to preserve the functionally important β1/β2 and the β3/β4 loops in separate subdomains. In addition, dissection at the β2/β3 loop seemed to cause the least steric effect

Gregory J. Miller (ed.), *Inositol Phosphates: Methods and Protocols*, Methods in Molecular Biology, vol. 2091,
https://doi.org/10.1007/978-1-0716-0167-9_4,

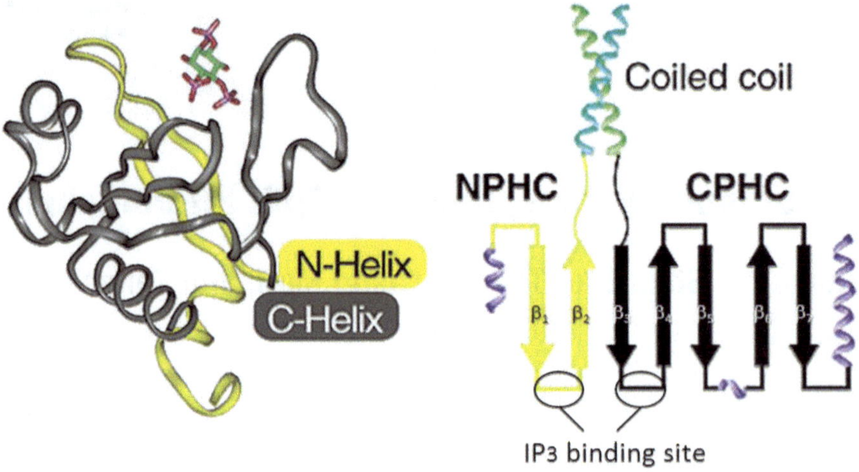

Fig. 1 The split PH domain designed from PLCδ$_1$ PH domain (PDB ID; 1MAI) [15]. (left) A ribbon representation shows a possible structure for NPHC (yellow) and CPHC (gray) reassembled into the PH domain. N-Helix and C-Helix denote the antiparallel coiled-coil segment of NPHC and CPHC, respectively. The IP$_3$ molecule found in the crystal structure of the complex of PLCδ$_1$ PH domain and IP$_3$ is shown in the stick representation (right). Arrows represent the β-sheets in the NPHC (yellow), and CPHC (gray). Oligopeptides forming an antiparallel coiled coil segment (coils in green) are attached at the C-terminal and N-terminal of NPHC and CPHC, respectively. (Figure modified from ref. 16)

when two subdomains were reassembled with an antiparallel coiled-coil formation. Isothermal titration microcalorimetry (ITC) measurements indicated a formation of a thermodynamically stable 1:1 complex of the N-terminal and C-terminal halves of the split PH domain by the coiled-coil formation. The reassembled split PH domain binds to D-Ins(1,4,5)P$_3$, a target molecule of the parent PLCδ$_1$ PH domain, with reasonable affinity but not to L-Ins(1,4,5) P$_3$, indicating that the split PH moiety maintains a binding selectivity similar to the native PLCδ$_1$ PH domain.

We further applied this methodology to reconstitute Bruton's tyrosine kinase (Btk) PH domain [17], which selectively binds Ins (1,3,4,5)P$_4$. Here, we have genetically inserted green fluorescent protein (GFP) in between the split PH domain [18] (Fig. 2). The design was conducted to transduce the ligand-induced subtle structural perturbation at the ligand binding site of PH domain to the chromophore of GFP. Using the split PH domain is expected to induce larger Ins(1,3,4,5)P$_4$-induced structural perturbation than the parent PH domain. By comparing the crystal structures before and after ligand binding, we realized the slight difference in the conformation of the loop region constituting the Ins(1,3,4,5)P$_4$ binding pocket. Therefore, this region was chosen to split the PH domain and insert the GFP moiety. With this design, we were able to specifically detect the subtle change of the PH domain structure accompanied by Ins(1,3,4,5)P$_4$ binding, which resulted in the change of the spectral property of the GFP moiety. One of

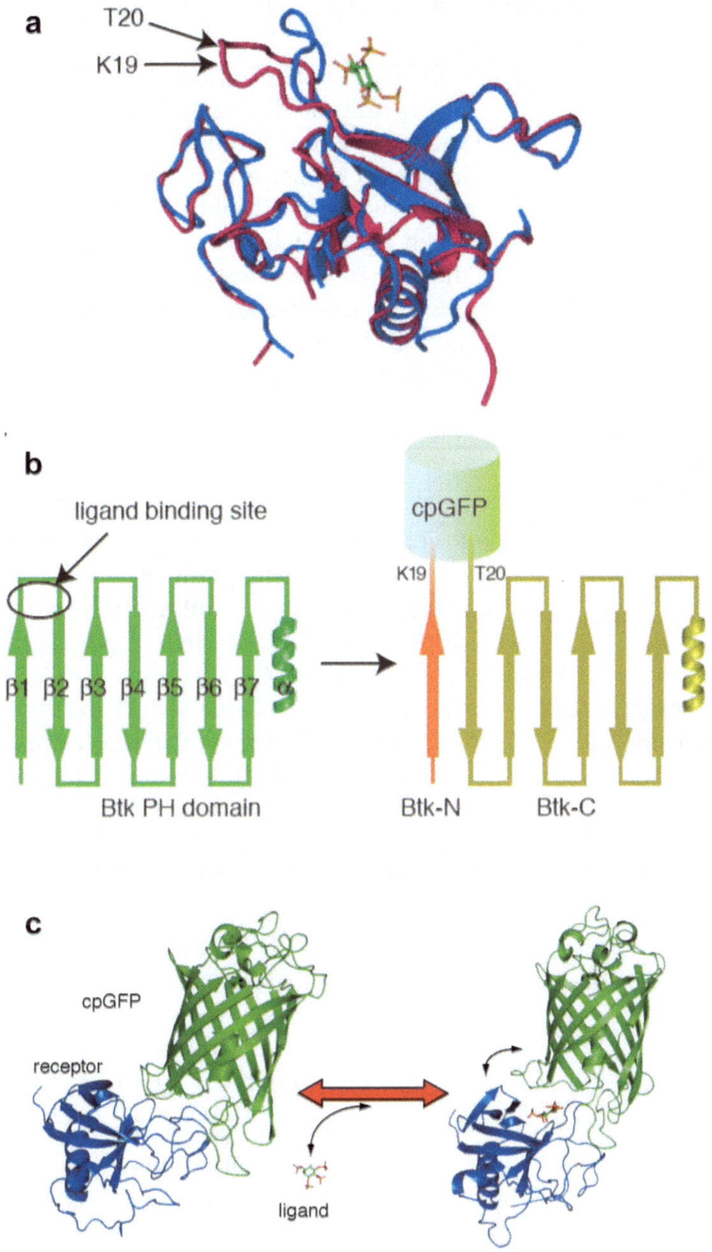

Fig. 2 (**a**) Superimposition of the three dimensional structures reveal only a slight structural alteration between the unliganded (magenta) and the Ins(1,3,4,5)P_4-bound (blue) states of Btk PH domain (PDB ID; 1B55, 1BTK). (**b**) Dissection strategy of Btk PH domain. Arrows represent the β-sheets of Btk PH domain (green), the Btk-N (orange) and Btk-C (ocher). The chromophore region of cpGFP is fused precisely to the ligand Ins(1,3,4,5)P_4 binding site. (**c**) Schematic illustration shows a fluorescent biosensor for Ins(1,3,4,5)P_4 based on the split Btk PH domain-cpGFP conjugate. The conformational change of the PH domain induced by the ligand-binding event was transduced to the structural change of the chromophore of cpGFP, and then resulted in the ratiometric fluorescence change of cpGFP. (Figure modified from ref. 18)

differences between the split PLCδ$_1$ PH domain is that the dividing site is right at the ligand binding region. Either of the constructs still maintains the binding property to the target ligand, manifesting the flexibility of the concept. These results demonstrate that the properly designed split PH domain folds into a functional structure when the split halves are brought to close proximity, and can be utilized to detect specific inositol phosphate of interest. Here, we describe the effective method to construct the properly designed split PH domains applied for PLCδ$_1$ PH domain and Btk PH domain.

2 Materials

Prepare all solutions using ultrapure water (i.e., milliQ or equivalent) and analytical grade reagents. Prepare and store all reagents at room temperature (unless indicated otherwise). Follow all waste disposal regulations when disposing of waste materials.

2.1 Design of Split PH Domain

1. Three-dimensional (3D) information of the PH domain of interest in a reasonable resolution. Information for both before and after substrate binding are required.

2.2 Construction of the Expression Vector

1. Primers and cDNA: Use the cDNA of your choice to amplify the gene encoding the PH domain of interest. The PCR primers can be purchased from various vendors. Desalted oligo is sufficient for purification.
2. DNA polymerase, restriction enzymes, ligase: DNA polymerase should be a high-fidelity type. Restriction enzymes should be chosen according to the sequence of the PH domain of interest so that the target gene would not be digested.
3. Gel extraction kit: Available from various vendors.
4. Competent cells: Two strains suited for cloning (such as DH5α) and protein expression (such as BL21(DE3)) will be necessary. Both are available from various vendors. Competent cells can also be prepared in the lab (*see* **Note 1**).
5. LB plates with appropriate antibiotics: Mix 10 g of Bacto tryptone, 5 g of yeast extract, 5 g of sodium chloride, 0.25 mL of 4 N NaOH, and 15 g of agar in dH_2O. Adjust the volume to 1 L. Autoclave the solution (121 °C, 20 min). Let the solution cool to ~50 °C, and add appropriate antibiotics (ampicillin, kanamycin, chloramphenicol, etc) depending on the plasmid and *E. coli* strain you are using. Pour the solution to petri dishes. Let it cool down well before use.
6. Mini-prep kit: Available from various vendors.

7. 50 × stock of TAE buffer for agarose gel: Dissolve 242 g of Tris-base, 57.1 mL of glacial acetic acid, 100 mL 0.5 M sodium EDTA in water and adjust the volume to 1 L. To make 1 × TAE from 50 × TAE buffer stock, dilute 20 mL of stock into 980 mL of dH_2O.
8. 1% Agarose gel: Measure 1 g of Electrophoresis grade agarose. Mix agarose powder with 100 mL 1 × TAE buffer in a beaker. Microwave for few min until the agarose is completely dissolved (*see* **Note 2**). Let agarose solution cool down to about 50 °C. Add ethidium bromide (EtBr) to a final concentration of 0.5 μg/mL. Pour the agarose into a gel tray with the well comb in place.

2.3 Protein Expression and Purification

1. IPTG (isopropyl-β-D-thiogalactopyranoside) stock (1 M): Dissolve 2.38 g of IPTG in dH_2O and adjust the volume of the solution to 10 mL. Sterilize by passing it through a 0.22-μm disposable filter. Dispense the solution into 1mL aliquots and store them at −20 °C.
2. PBS and PBS containing 4 M urea (for insoluble proteins). Add protease inhibitor (such as cOmplete™ from Roche) just before use.
3. Ion exchange chromatography buffers (PBS and PBS containing 1 M NaCl, pH 7.4). Solutions should be filtered through 0.22 μm filter and degassed well before use.

2.4 SDS-PAGE

1. Resolving gel buffer: 1.5 M Tris–HCl, pH 8.8. Adjust pH with HCl. Store at 4 °C.
2. Stacking gel buffer: 0.5 M Tris–HCl, pH 6.8. Adjust pH with HCl. Store at 4 °C.
3. Thirty percent acrylamide/Bis solution (29:1): Dissolve 29.0 g of acrylamide monomer and 1.0 g *N*,*N*-′-methylenebisacrylamide (cross-linker) and transfer to a 100 mL graduated cylinder. Adjust volume to 100 mL with dH_2O and filter through a 0.45 μm filter. Store at 4 °C in a light-protective bottle (*see* **Note 3**).
4. Ammonium persulfate: 10% solution in dH_2O.
5. *N*,*N*,*N*′,*N*′-Tetramethylethylenediamine (TEMED).
6. SDS-PAGE running buffer: 0.025 M Tris–HCl, pH 8.3, 0.192 M glycine, 0.1% SDS (*see* **Note 4**)
7. Staining solution: 0.1% Coomassie Brilliant Blue R-250, 50% methanol and 10% glacial acetic acid in dH_2O.
8. Destaining solution: 40% methanol and 10% glacial acetic acid in dH_2O.

2.5 Functional Analysis of the Split PH Domain

1. Target inositol phosphate solutions: Dissolve the entire bottle/vial of the target inositol phosphate (because typically the amount is too small to weigh accurately) in the buffer of your choice, such as PBS or 10 mM Tris–HCl (pH 7.4) (*see* **Note 5**). Aliquot the solution and store in −20 °C. Avoid freeze-thaw.

3 Methods

3.1 Design of Split PH Domain Based on 3D Structure

1. Fusing the dimerization module: An appropriate 3D structural information of the PH domain of interest, typically found in the protein data bank (Fig. 1). Usually, it is not difficult to locate the dissection position for subunits near the binding site to inositol phosphate derivatives. Several candidate residues for the dissection site should be considered. Care should be taken not to split at the functionally important motif, such as the β1/β2 and the β3/β4 loops of PLCδ_1 PH domain, to preserve the ligand binding affinity. The split PH domain can be fused with a dimerization module (such as a coiled-coil motif). In the case of PLCδ_1 PH domain, the PH domain was divided into two subunits at N45 and K49.
2. Insertion to GFP: In case the structural information of not only the substrate-unbound state but also the substrate–bound state of the PH domain of your interest are obtained, the region responsible for the conformational change associated with the ligand binding event would be deduced from the comparison of both structures. Superimposition of the free and the ligand-bound structures would visualize the conformationally altered regions (Fig. 2a). Several candidate residues for the dissection site would be listed. The residue near the ligand binding site would transduce structural disturbance effectively, but note that residues directly interacting with the bound ligand should be avoided to maintain the selectivity and affinity of the split PH domain. In the case of Btk PH domain, by following this method, K19 and T20 were selected as the dissection point for the split PH domain to insert cpGFP (Fig. 2).

3.2 Construction of the Expression Vector

1. Amplification of the target gene encoding the PH domain: Design DNA oligos that can amplify the two halves of split PH domain. If you would like to fuse some modification such as a coiled-coil domain, prepare the oligos accordingly. The oligos should contain restriction enzyme sites for cloning and a stop codon (if applicable). Use the cDNA that expresses the PH domain of your interest as a template to amplify the target gene. DNA polymerase should be high-fidelity type. Run the product on 1% agarose gel and cut out the band of the expected size to extract the gene of interest.

2. Insertion of the gene of interest into an expression vector: Digest both the vector and the amplified gene of interest with appropriate restriction enzyme. Dephosphorylate the vector using alkaline phosphatase. Ligate the gene of interest into the vector by mixing them at the molar ratio of 3–10:1. Total DNA amount should be 0.1–1 g. Follow the manufacturer's instruction for ligation, typically incubating the mixture with DNA ligase at 16 °C for 30–60 min.
3. Transformation to *E. coli*: Take competent cells out of −80 °C and thaw on ice. Take agar plate containing the appropriate antibiotics from storage at 4 °C and let it warm to room temperature. Mix the ligated DNA mixture into ~50 μL of competent cells. Gently mix the tube. Incubate the competent cell–DNA mixture on ice for 20–30 min. Heat-shock the tube by placing it a heat block at 42 °C water bath for 30–60 s. Put the tubes back on ice for 2 min. Add 1000 μL LB medium (without antibiotics) to the bacteria and grow in 37 °C shaking incubator for 45 min. Plate the transformation solution onto a LB agar plate containing the appropriate antibiotic.
4. Plasmid extraction/sequence confirmation: Pick up 4–10 colonies from the plate and culture in 5 mL LB medium each. Extract the plasmids with a mini-prep kit of your choice and confirm the DNA sequence.

3.3 Protein Purification

1. Protein expression in *E. coli* expression system: Transform the expression vector to an *E. coli* strain for protein expression (such as BL21 (DE3) or Rosetta 2). Pick up a colony and preculture in 5 mL LB medium. Transfer the culture to 1 L of LB medium (with appropriate antibiotics) and monitor the OD_{600} until 0.4~0.5. Add IPTG to a final concentration of 0.5 mM to induce protein expression. Continue the culture for 4 h at 37 °C. The culture volume, IPTG concentration, and culture time and temperature can be optimized depending on the characteristics of the protein.
2. Sonication: Spin down the *E. coli* culture to obtain the pellet. Resuspend the pellet by PBS. Transfer the pellet solution to 40 mL centrifuge tube. Add protease inhibitor if possible. Put the centrifuge tube on ice. Perform sonication by output 20–30%, 1 s ON/1 s OFF, total run 10–12 min. Adjust power and time according to the amount of sample. Make sure not to increase the sample temperature too much during the sonication step. Spin down the lysate to separate the supernatant and the pellet. To clarify the presence of the protein of interest, supernatant and pellet should be checked by SDS-PAGE. Occasionally, the protein of interest might be included in the pellet. In this case, discard the supernatant and resuspend the pellet by PBS containing 4 M urea. Briefly

sonicate the solution and spin down again to recover the supernatant.

3. Ion exchange chromatography: Filter the sample solution through 0.22 μm filter. Because PH domains are typically basic, cation exchange column can be used. We typically use HiTrap SP for rough separation and MonoS for last step purification (both from GE Healthcare Life Sciences). Load the sample solution to the column. Set the gradient from 0% to 100% solution B (PBS containing 1 M NaCl, pH 7.4) and collect the fractions. Flow rate and length of gradient can be adjusted according to the column. Determine the fraction containing the protein of interest and go through the second column for last step purification.
4. Analysis by 15% SDS-PAGE: Mix 2.5 mL of resolving buffer, 5.0 mL of acrylamide mixture, and 2.3 mL water in a 50 mL Falcon tube. Add 100 μL of 10% SDS, 80 μL of ammonium persulfate, and 10 μL of TEMED and cast the gel in a gel cassette. Let it sit for at least 1 h. Prepare the stacking gel by mixing 1.25 mL of stacking buffer, 0.67 mL of acrylamide mixture, and 3 mL water in a 50 mL Falcon tube. Add 100 μL of 10% SDS, 40 μL of ammonium persulfate, and 5 μL of TEMED. Pour the mixture on top of the resolving gel and immediately insert a gel comb. Add loading buffer to the samples and heat them at 95 °C for 5 min. Perform electrophoreses at 10 mA/gel until the sample has entered the resolving gel and then continue at 25 mA/gel until the BPB dye front reaches the bottom of the gel. Gently take out the gel from the glass gel cassette and stain the gel in CBB staining solution for 20–30 min with gentle agitation, followed by destaining of the gel in destaining solution until the background of the gel is completely destained. Determine the purity of the sample by the gel image (Fig. 3).

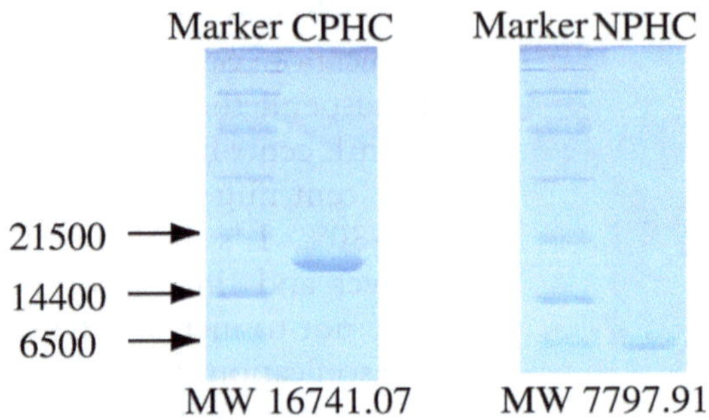

Fig. 3 The purity and molecular weight of C-terminal half of PLCδ_1 PH domain fused with coiled-coil (CPHC, pI = 9.36) and N-terminal half of PLCδ_1 PH domain fused with coiled-coil (NPHC, pI = 6.24) were confirmed by 15% SDS/polyacrylamide gel. (Figure modified from ref. 16)

3.4 Functional Analysis of the Split PH Domain

1. ITC (MicroCal iTC200): Make sure that the samples are purified well to avoid detecting background (*see* **Note 6**). Adjust the concentration of PH domain solution so that sufficient amount of heat would be produced. Set the PH domain solution to the sample chamber; place the target inositol phosphate solution in the titration sample syringe. Titrate the protein solution by target inositol phosphate to measure the heat production accompanied with substrate binding. Determine the affinity and specificity against the target inositol phosphate (Fig. 4).
2. Fluorescence spectrometer (Hitachi F-7000): For GFP-fusion type PH domain, substrate binding can also be monitored by fluorescent properties. Dilute the purified PH domain. Measure the fluorescent spectra without substrate. Add the target inositol phosphate and determine the difference of the fluorescent spectra. Continuously add the target inositol phosphate and measure the spectrum to determine the saturation point (Fig. 5). Calculate the K_d by plotting the difference of the

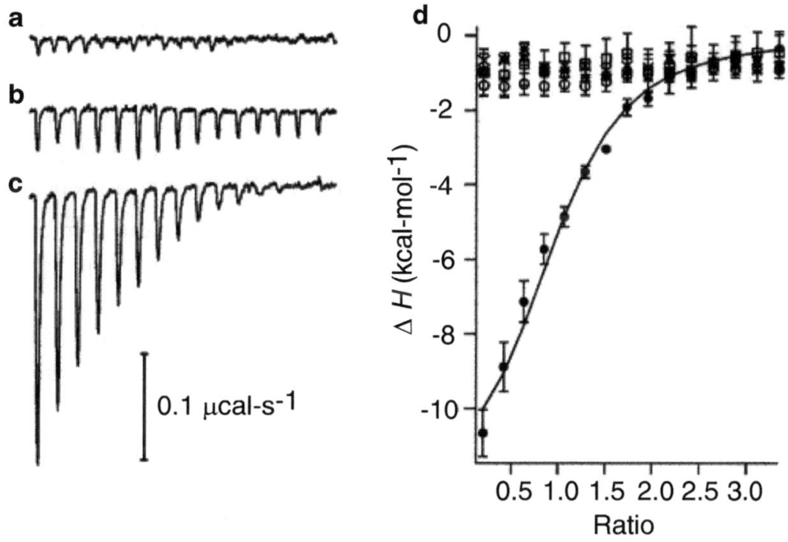

Fig. 4 Titration of NPHC/CPHC complex with L-Ins(1,4,5)P_3 and D-Ins(1,4,5)P_3 by ITC. Titrations of the 1:1 NPH/CPH (N and C terminus half of PLCδ$_1$ PH domain without the coiled-coil, respectively) mixture (4 µM) with 80 µM D-Ins(1,4,5)P_3 (**a**), and NPHC/CPHC complex (4 µM) with 80 µM L-Ins(1,4,5)P_3 (**b**) or D-Ins (1,4,5)P_3 (**c**) added in 15 aliquots of 15 µL each. The heat of dilution was subtracted from the raw data prior to data analysis. (**d**) Analyzed data for isothermal titration of the NPHC/CPHC complex with L-Ins(1,4,5)P_3 (open circles) or D-Ins(1,4,5)P_3 (filled circles) indicate a specific D-Ins(1,4,5)P_3 binding of the NPHC/CPHC complex. (Figure modified from ref. [16])

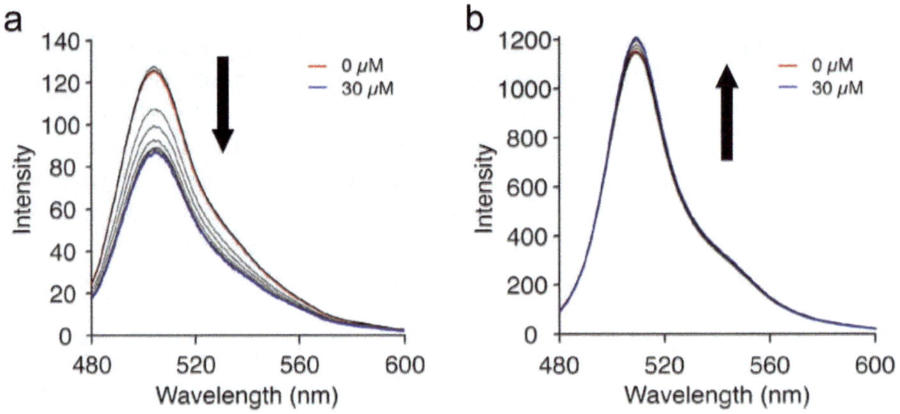

Fig. 5 Fluorescence spectral changes of Btk-cpGFP ((**a**) excited at 470 nm, (**b**) excited at 396 nm) upon addition of Ins(1,3,4,5)P_4. (Figure modified from ref. 18)

fluorescent intensity against substrate concentration. Repeat this using other inositol phosphates to evaluate the specificity against the target inositol phosphate.

4 Notes

1. For some proteins with low expression level (possessing rare codons or toxicity), engineered *E. coli* strains such as Rosetta™ 2 or Tuner (DE3) are recommended for better yield.
2. It is better to take out and stir the solution once in a while during the microwave heating to assure the agarose to melt evenly.
3. Monomer acrylamide is toxic. Make sure to use appropriate Personal Protective Equipment such as lab coats, nitrile gloves, safety glasses, and masks.
4. SDS running buffer can be reused 2–3 times. Further repeated usage is not recommended because the stacking efficiency of the protein becomes low.
5. It is easier to directly put the exact amount of buffer (calculated to obtain the desired concentration of inositol phosphate) into the vial and dissolve the entire contents. Typical stock solutions of inositol phosphates are 1–5 mM.
6. Dialyze the protein against the measurement buffer well, because mixing of buffers with different components causes background heat. Put the protein solution in a dialysis tube and dialyze against 1000-fold volume of the buffer three times. Ideally the buffer should be the same as the solution used to dissolve the target inositol phosphate.

References

1. Gibson TJ, Hyvonen M, Musacchio A, Saraste M, Birney E (1994) PH domain: the first anniversary. Trends Biochem Sci 19:349–353
2. Walliser C, Retlich M, Harris R, Everett KL, Josephs MB, Vatter P, Esposito D, Driscoll PC, Katan M, Gierschik P, Bunney TD (2008) rac regulates its effector phospholipase Cgamma2 through interaction with a split pleckstrin homology domain. J Biol Chem 283:30351–30362
3. van Rossum DB, Patterson RL, Sharma S, Barrow RK, Kornberg M, Gill DL, Snyder SH (2005) Phospholipase Cgamma1 controls surface expression of TRPC3 through an intermolecular PH domain. Nature 434:99–104
4. Yan J, Wen W, Xu W, Long JF, Adams ME, Froehner SC, Zhang M (2005) Structure of the split PH domain and distinct lipid-binding properties of the PH-PDZ supramodule of alpha-syntrophin. EMBO J 24:3985–3995
5. Wen W, Liu W, Yan J, Zhang M (2008) Structure basis and unconventional lipid membrane binding properties of the PH-C1 tandem of rho kinases. J Biol Chem 283:26263–26273
6. Lu Q, Yu J, Yan J, Wei Z, Zhang M (2011) Structural basis of the myosin X PH1(N)-PH2-PH1(C) tandem as a specific and acute cellular PI(3,4,5)P(3) sensor. Mol Biol Cell 22:4268–4278
7. Teo H, Gill DJ, Sun J, Perisic O, Veprintsev DB, Vallis Y, Emr SD, Williams RL (2006) ESCRT-I core and ESCRT-II GLUE domain structures reveal role for GLUE in linking to ESCRT-I and membranes. Cell 125:99–111
8. Ahn AH, Freener CA, Gussoni E, Yoshida M, Ozawa E, Kunkel LM (1996) The three human syntrophin genes are expressed in diverse tissues, have distinct chromosomal locations, and each bind to dystrophin and its relatives. J Biol Chem 271:2724–2730
9. Lemmon MA, Ferguson KM, Schlessinger J (1996) PH domains: diverse sequences with a common fold recruit signaling molecules to the cell surface. Cell 85:621–624
10. Falasca M, Logan SK, Lehto VP, Baccante G, Lemmon MA, Schlessinger J (1998) Activation of phospholipase C gamma by PI 3-kinase-induced PH domain-mediated membrane targeting. EMBO J 17:414–422
11. Iwata Y, Pan Y, Yoshida T, Hanada H, Shigekawa M (1998) Alpha1-syntrophin has distinct binding sites for actin and calmodulin. FEBS Lett 423:173–177
12. Rhee SG (2001) Regulation of phosphoinositide-specific phospholipase C. Annu Rev Biochem 70:281–312
13. Chang JS, Seok H, Kwon TK, Min DS, Ahn BH, Lee YH, Suh JW, Kim JW, Iwashita S, Omori A, Ichinose S, Numata O, Seo JK, Oh YS, Suh PG (2002) Interaction of elongation factor-1alpha and pleckstrin homology domain of phospholipase C-gamma 1 with activating its activity. J Biol Chem 277:19697–19702
14. Scheffzek K, Welti S (2012) Pleckstrin homology (PH) like domains - versatile modules in protein-protein interaction platforms. FEBS Lett 586:2662–2673
15. Ferguson KM, Lemmon MA, Schlessinger J, Sigler PB (1995) Structure of the high affinity complex of inositol trisphosphate with a phospholipase C pleckstrin homology domain. Cell 83:1037–1046
16. Sugimoto K, Mori Y, Makino K, Ohkubo K, Morii T (2003) Functional reassembly of a split PH domain. J Am Chem Soc 125:5000–5004
17. Baraldi E, Djinovic Carugo K, Hyvonen M, Surdo PL, Riley AM, Potter BV, O'Brien R, Ladbury JE, Saraste M (1999) Structure of the PH domain from Bruton's tyrosine kinase in complex with inositol 1,3,4,5-tetrakisphosphate. Structure 7:449–460
18. Sakaguchi R, Endoh T, Yamamoto S, Tainaka K, Sugimoto K, Fujieda N, Kiyonaka S, Mori Y, Morii T (2009) A single circularly permuted GFP sensor for inositol-1,3,4,5-tetrakisphosphate based on a split PH domain. Bioorg Med Chem 17:7381–7386

Chapter 5

Dictyostelium discoideum as a Model to Study Inositol Polyphosphates and Inorganic Polyphosphate

Yann Desfougères and Adolfo Saiardi

Abstract

The yeast *Saccharomyces cerevisiae* has given us much information on the metabolism and function of inositol polyphosphates and inorganic polyphosphate. To expand our knowledge of the metabolic as well as functional connections between inositol polyphosphates and inorganic polyphosphate, we have refined and developed techniques to extract and analyze these molecules in a second eukaryotic experimental model, the amoeba *Dictyostelium discoideum*. This amoeba, possessing a well-defined developmental program, is ideal to study physiological changes in the levels of inositol polyphosphates and inorganic polyphosphate, since levels of both molecules increase at late stages of development. We detail here the methods used to extract inositol polyphosphates using perchloric acid and inorganic polyphosphate using acidic phenol. We also present the postextraction procedures to visualize and quantify these molecules by polyacrylamide gel electrophoresis and by malachite green assay.

Key words Inositol polyphosphate, Extraction, DAPI staining, Toluidine blue, Electrophoresis, polyP, Inositol pyrophosphate, Phytic acid

1 Introduction

The linear polymer of phosphate groups (PO_4^{3-}) called inorganic polyphosphate (polyP) is primarily studied in bacteria, but its presence in eukaryotes is well documented [1, 2]. Conversely, while inositol can be synthesized in bacteria [3], only nucleated cells have exploited the inositol ring to generate inositol polyphosphates (IPs). These are molecules with differentially arranged phosphate groups, attached to inositol through phosphoester bonds [4]. Furthermore, inositol can also contain diphosphate moieties, that is, highly energetic phosphoanhydride bonds; these species are called inositol pyrophosphates (PP-IPs) [5, 6]. Recent studies primarily executed in the yeast *Saccharomyces cerevisiae* have demonstrated that the metabolism of polyP and PP-IPs is interconnected [7, 8]. This yeast has been instrumental in the elucidation of the inositol polyphosphate metabolic pathway, which from

Gregory J. Miller (ed.), *Inositol Phosphates: Methods and Protocols*, Methods in Molecular Biology, vol. 2091, https://doi.org/10.1007/978-1-0716-0167-9_5,

phospholipase C-generated Ins(1,4,5)P_3 produces the fully phosphorylated IP_6 (inositol hexakisphosphate or phytic acid) and the PP-IPs IP_7 and IP_8 (diphosphoinositol pentakisphosphate and bis-diphosphoinositol tetrakisphosphate) [9–13]. Similarly, yeast has been fundamental to the identification of many physiological functions attributed to IPs and PP-IPs, such as vesicular trafficking and telomere length [14–16]. The studies of *S. cerevisiae* polyP metabolism have been facilitated by the accumulation of this polymer in the yeast vacuole and thus by relatively simple extraction and analysis procedures. The vacuolar transporter chaperone (VTC) complex, specifically the subunit Vtc4, is responsible for polyP synthesis in yeast; a recent work revealed that the IPs/PP-IPs regulate polyP synthesis by binding to the SPX domain of VTC proteins [8].

While VTC homologous proteins are present in other eukaryote clades such as the kinetoplastida (trypanosomes) [17], they are absent from the majority of eukaryotes. In another eukaryotic organism, the social amoeba *Dictyostelium discoideum*, polyP is synthesized by polyphosphate kinase (DdPPK or simply PPK), an enzyme of bacterial origin obtained by horizontal gene transfer [18]. In *D. discoideum* polyP is important for general fitness, since amoebae lacking this polymer grow slowly [19]. Interestingly, polyP dramatically accumulates during the amoeba's development and is stored in the spores. It has also been reported that polyP is somehow secreted into the culture medium. The absence of polyP accumulation during the development of ppk1 mutant cells is offset by an increase of both ATP and PP-IPs, suggesting that these three phosphate-rich molecules interactively balance each other [19].

While *D. discoideum* has become a polyP experimental model only recently, the amoeba has been instrumental for the initial discovery, during the 1980s and early 1990s, of many IPs and their metabolism [20–22]. Furthermore, it was *D. discoideum* where the PP-IPs were originally discovered [23]. These discoveries were facilitated by the amoeba's high levels of IP_6, IP_7, and IP_8. These highly phosphorylated IP species stain very well with toluidine blue once resolved by polyacrylamide gel electrophoresis (PAGE) [24]. Therefore, PAGE analysis of *D. discoideum* cell extracts easily reveals PP-IPs species and opens new investigative opportunities [25]. Interestingly, polyP analysis is similarly performed by toluidine staining or alternatively by DAPI staining once resolved by PAGE [26]. Thus these two phosphate-rich families of molecules share similar extraction and analysis protocols, although slightly different procedures are required to optimize the analysis of polyP or PP-IPs.

Here, we present the optimal protocol to extract and analyze by PAGE both polyP, IP_6 and PP-IPs from *D. discoideum*. Perchloric acid extraction is preferred to purify small acidic molecules such as IPs, while phenol extraction is better suited to purify the larger polyP polymer. A high acrylamide concentration is required for PAGE to resolve IPs, which stain strongly with toluidine blue (Fig. 1), while polyP can additionally be visualized by negative DAPI staining (Fig. 2). We also discuss the classical malachite green assay to quantify polyP once it is degraded to phosphate (Fig. 3).

2 Materials

2.1 Cell Culture

HL5 medium (per liter): 14 g peptone, 7 g yeast extract, 13.5 g glucose, 0.5 g KH_2PO_4, 0.5 g Na_2HPO_4, filter-sterilized or autoclaved. Store at 4 °C (*see* **Note 1**).

KK2 buffer: 20 mM KH_2PO_4/K_2HPO_4 pH 6.8, filter sterilized or autoclaved. Store at 4 °C.

KK2 plates: 1.5% agar in KK2 buffer, autoclaved. Store at 4 °C.

The *D. discoideum* strains used, namely, wild type axenic strain (AX2) and the strains deleted for ppkl and IP_6-kinase (i6ka), are available through the DictyBase Stock Center (http://dictybase.org).

2.2 Extraction of Inositol Polyphosphates

Perchloric acid 1 M (*see* **Note 2**).

Potassium carbonate 1 M (*see* **Note 2**).

Ethylenediaminetetraacetic acid (EDTA) 0.5 M pH 8.0 (Invitrogen #15575-038).

2.3 Extraction of Inorganic Polyphosphate

LETS buffer: 0.1 M LiCl; 0.01 M Na_2EDTA; 0.01 M Tris–Cl pH 7.4; 0.2% SDS.

Phenol, saturated with 0.1 M citrate buffer, pH 4.3. Store at 4 °C.

Chloroform.

Ethanol, both pure and 70% solution in water.

10× Tris–Borate–EDTA buffer (TBE): 450 mM Tris–borate, 10 mM EDTA.

2.4 Polyacrylamide Gel Electrophoresis (PAGE)

40% polyacrylamide (19:1 acrylamide–bis-acrylamide).

Ammonium persulfate, 10% solution in water. Store at −20 °C.

N,*N*,*N*,*N'*-Tetramethyl-ethylenediamine (TEMED).

polyP standards, 10 mM in water. Store at 4 °C (*see* **Note 3**).

IP_6 standard, 1 mM in water. Store at 4 °C (*see* **Note 4**).

Orange G.

Loading buffer: 50% glycerol, 0.01% Orange G, 1 mM EDTA.

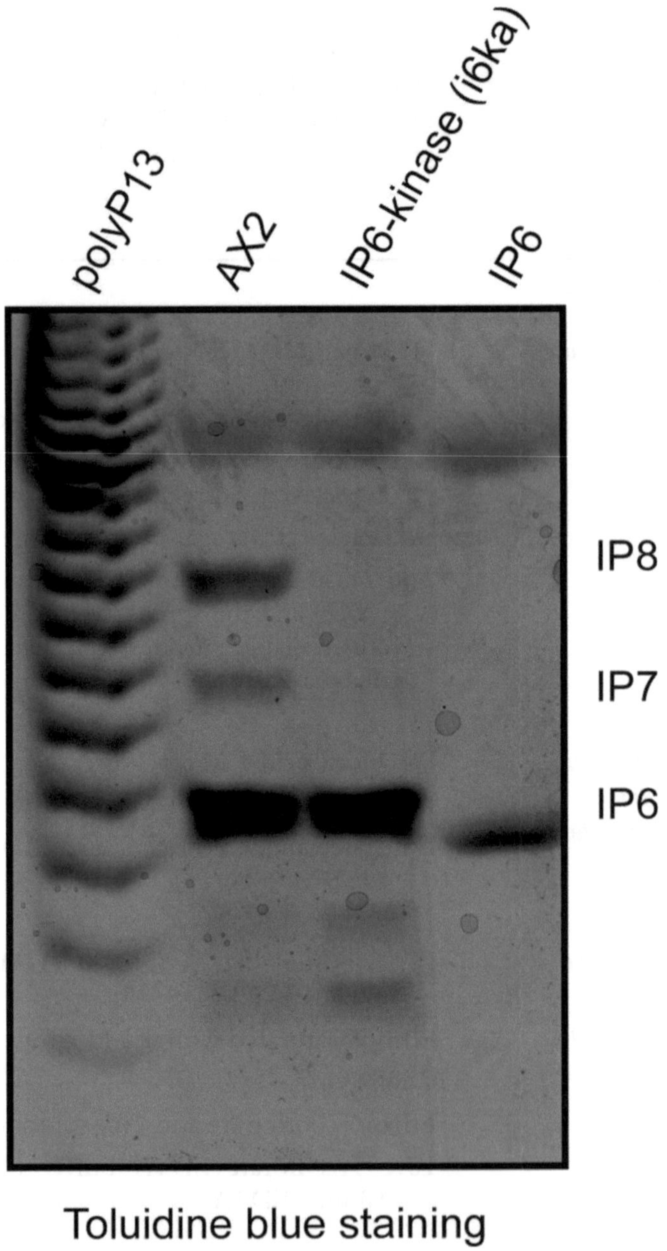

Fig. 1 Representative IP profiles of *D. discoideum* AX2 and IP_6-kinase knockout (i6ka). Inositol polyphosphates extracted from 2×10^7 cells with perchloric acid were loaded onto a 35.5% polyacrylamide gel. Inositol hexakisphosphate (IP_6) and polyphosphate of mean chain length of 13 phosphate residues ($polyP_{13}$) were loaded as standards. The Orange G dye was used to follow the gel migration. The gel was stained with toluidine blue, scanned, and processed using Adobe Photoshop

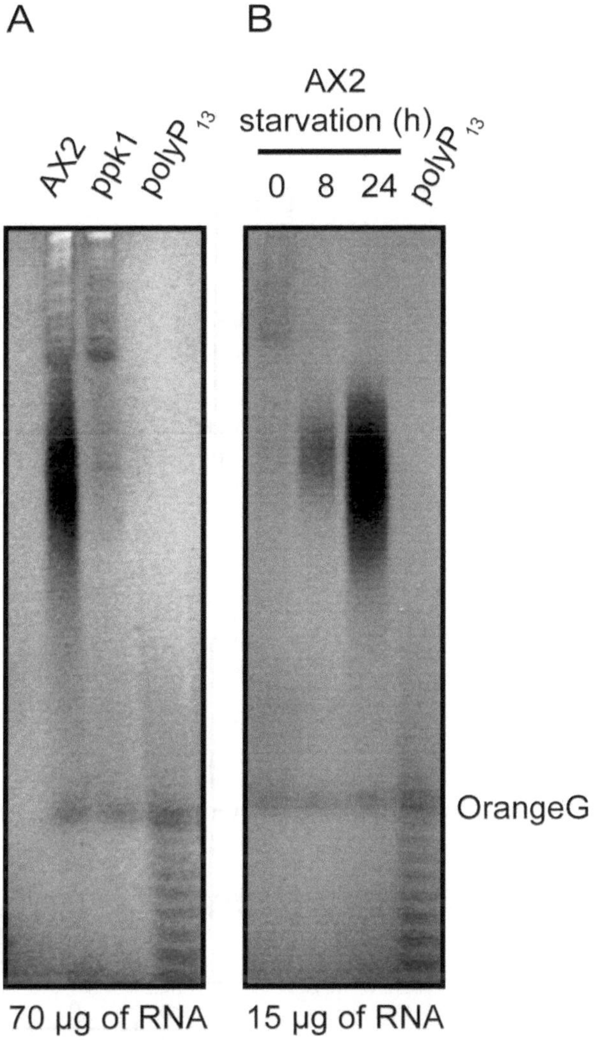

Fig. 2 polyP profiles of *D. discoideum* AX2 and polyphosphate kinase knockout (ppk1). The polyP polymer was extracted using the acidic phenol protocol and loaded onto a 25% polyacrylamide gel. Orange G was used to follow the gel migration, and $polyP_{13}$ was used as a standard. The gels were stained with DAPI; polyP is visualized as negative staining since it induces DAPI photobleaching. (**a**) The polyP polymer (70 μg of RNA) was extracted from vegetative growing cells and resolved on gel. (**b**) Gel resolving polyP (15 μg of RNA) extracted from cells collected at different times during development on KK2

2.5 Toluidine Blue Staining

Toluidine blue staining solution (0.1% Toluidine blue; 20% Methanol; 2% Glycerol) (*see* **Note 5**).

Photo Scanner (Epson Perfection 4990 PHOTO or similar equipment).

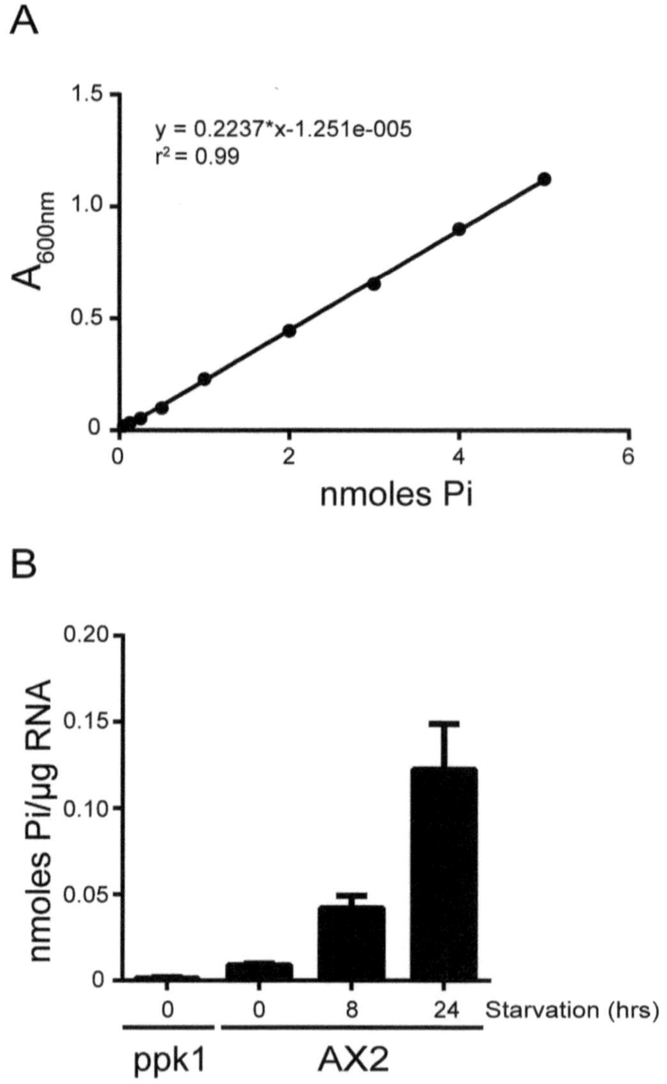

Fig. 3 Quantification of polyP using the malachite green assay. (**a**) Representative standard curve obtained with 0.05 to 5 nmoles of free phosphate (Pi) KH_2PO_4. (**b**) Purified polyP was quantified as the amount of Pi released after the action of the recombinant polyphosphatases Ppx1 and Ddp1

2.6 DAPI Staining

DAPI stock solution (10 mg/ml DAPI in water)

Destaining solution (20% Methanol; 2% Glycerol).

UV 300 nm transillumination recording chamber.

2.7 Malachite Green Assay of Inorganic Phosphate

Molybdate solution: 28 mM ammonium heptamolybdate (Fluka #09880) in 2.1 M sulfuric acid.

Malachite green solution: 0.76 mM malachite green (Sigma-Aldrich #213020) in 0.35% polyvinyl alcohol. Store at 4 °C (*see* **Note 6**).

Polyphosphatase buffer (5 mM Tris–HCl pH 7.0, 1.5 mM $MgCl_2$).

Recombinantly produced His-ScDdp1 and His-ScPpx1. Store at −20 °C (*see* **Note 7**).

Absorbance spectrophotometer or plate reader.

3 Methods

3.1 D. discoideum Growth and Starvation Procedures

1. Grow *D. discoideum* cells axenically in HL5 media at 22 °C (120 rpm) to a cell density not exceeding 4×10^6 cells/ml, ideally between $1\text{–}3 \times 10^6$ cells/ml (*see* **Note 8**).
2. To induce development, collect 10^7 cells at a cell density of $1\text{–}4 \times 10^6$ cells/ml.
3. Wash twice in KK2 buffer and plate onto 55 mm diameter KK2 plates.
4. After 8 h or 24 h, collect the cells manually by adding 500 μl of KK2 buffer to the plates.
5. Scrape the cells out of the plates using a plastic spatula.
6. Transfer the cell suspension to a 1.5 ml microtube.
7. Wash the plate once more as before and pool the cells.
8. Wash the cells once in KK2 buffer.

3.2 Inositol Polyphosphates Extraction

1. Grow the cells to a density of $1\text{–}4 \times 10^6$ cells/ml.
2. Centrifuge $0.5\text{–}2 \times 10^7$ cells ($1000 \times g$, 2 min, 22 °C).
3. Wash the cells once with KK2 buffer and centrifuge as before.
4. Resuspend the cell pellets in 2 volumes or at least 50 μl of 1 M perchloric acid containing 5 mM EDTA. Prepare freshly (*see* **Note 2**).
5. Vortex at 4 °C for 10 min.
6. Centrifuge at $16{,}000 \times g$ for 5 min at 4 °C.
7. Transfer the supernatant to a new microtube.
8. Add 1 volume of 1 M potassium carbonate containing 5 mM EDTA. Prepare freshly (*see* **Note 2**) (*see* **Note 9**).
9. Leave the tubes on ice for at least 2 h with occasional shaking and the lid open.
10. Centrifuge at $16{,}000 \times g$ for 5 min at 4 °C.
11. The recovered supernatant contains the inositol polyphosphates (*see* **Note 10**).

3.3 Extraction of polyP

1. Grow the cells to a density of $1\text{–}4 \times 10^6$ cells/ml.
2. Centrifuge $2\text{–}4 \times 10^7$ cells ($1000 \times g$, 2 min, 22 °C).
3. Wash the cells once with KK2 buffer and centrifuge as before.

4. Resuspend the cells pellet in 1.5–2 volumes of LETS buffer, or at least 200 μl.
5. Add one volume of phenol (pH 4.3) to the cell suspension.
6. Vortex the samples at 4 °C for 5 min.
7. Centrifuge the samples at 16,000 × *g* for 5 min at 4 °C.
8. Transfer the aqueous top layer to a new 1.5 ml microtube.
9. Add 2 volumes of chloroform and mix vigorously.
10. Vortex the samples at 4 °C for 5 min.
11. Centrifuge at 5000 × *g* for 5 min at 4 °C.
12. Transfer the aqueous phase to a new 1.5 ml microtube.
13. Add 2.5 volumes of ethanol.
14. Keep the samples at −80 °C for 2 h or overnight.
15. Centrifuge the samples at 16,000 × *g* for 5 min at 4 °C.
16. Wash the pellets once with 70% ethanol.
17. Resuspend the pellets in water or 1× TBE.
18. Measure RNA concentration by recording the absorbance at 260 nm (*see* **Note 11**).

3.4 Inositol Phosphates Analysis by PAGE and Toluidine Blue Staining

1. Prepare 38 ml of gel solution by mixing polyacrylamide–bis-acrylamide stock (40%), TBE and water to obtain a final polyacrylamide concentration of 33–35% (*see* **Note 12**).
2. Add 200 μl of 10% APS and 20 μl of TEMED just prior to pouring the gel.
3. Pour the gel slowly, avoiding the formation of air bubbles, and fit the comb.
4. Wait for the gel to set (around 20–30 min) at room temperature (*see* **Note 13**).
5. Allow the gel to set completely in the cold room to prevent it from drying.
6. Prerun the gel at 3 mA for 30–60 min.
7. Add sample buffer to the samples to be analyzed for a final glycerol concentration of 10%.
8. Load the samples onto the gel.
9. Run the gel in the cold room at 500–600 V and 5–6 mA for around 16–18 h (*see* **Note 14**).
10. After the dye reaches two thirds of the gel, turn off the power supply (*see* **Note 14**).
11. Remove the gel from the plates and transfer it to a plastic box.
12. Add the toluidine blue staining solution to the box.

13. Incubate the gel at room temperature for 30 min with gentle shaking.
14. Recover the staining solution (*see* **Note 15**).
15. Rinse the gel with 20% methanol.
16. Destain the gel in 20% methanol until the desired staining is obtained changing the 20% methanol every 30 min.
17. Scan the gel. The typical result of this analysis when performed on AX2 and IP_6-kinase null cells (i6ka) is presented in Fig. 1.

3.5 PAGE Analysis of polyP and DAPI Staining

1. The same procedure as in Subheading 3.3 must be followed to prepare the gels, except that lower polyacrylamide concentrations (between 20% and 30% acrylamide) should be used (*see* **Note 16**).
2. Run the gel in the cold room at 400–500 V and 4–5 mA for around 16–18 h (*see* **Note 14**).
3. Pour one about liter of 20% methanol, 2% glycerol solution containing 2 g of Tris base (pH>8.0) into a plastic box.
4. Add one 200 μl aliquot of 10 mg/ml DAPI and mix to complete the staining solution.
5. Remove the gel from the apparatus and incubate it in the staining solution for 40 min with gentle shaking and illumination.
6. Pour off the staining solution and rinse the gel twice with 250 ml of destaining solution.
7. Destain for 30 min with gentle shaking.
8. Use the UV transilluminator to image the gel (*see* **Note 15**). The typical result of this analysis when performed on AX2 and ppk1 null cells is presented in Fig. 2.

3.6 Quantification of polyP Using the Malachite Green Detection Assay

1. PolyP is extracted as described in Subheading 3.3.
2. Prepare serial dilutions of polyP samples in polyphosphatase buffer to a final volume of 100 μl.
3. Split the sample into two wells of a 96-well plate and add 0.01 μg/μl of recombinant ScPpx1 and ScDdp1 to one of the aliquots.
4. In the meantime, prepare serial dilutions of KH_2PO_4 from 0.05 to 5 nmoles in polyphosphatase buffer.
5. Add 100 μl of each dilution into the 96-well plate.
6. Incubate at 37 °C for 1 h.
7. Prepare the malachite green reagent just before use by mixing 860 μl of molybdate solution and 640 μl of malachite green solution.
8. Add 150 μl of malachite green reagent to all the samples.

9. Incubate the plate at room temperature for 5 min.
10. Read the absorbance at 595–600 nm.
11. Express the amount of free phosphate released by His-ScPpx1/His-ScDdp1 relative to the initial amount of RNA in the sample (nmole Pi/μg of RNA), as represented in Fig. 3.

4 Notes

1. All solutions are prepared using milliQ or equivalent ultrapure water (18 MΩ-cm); all solutions and or reagents are stored at room temperature (unless otherwise indicated).
2. EDTA (0.5 mM) is freshly added to the required amount (a few ml) of perchloric acid and potassium carbonate 1 M stock solutions. These prepared aliquots can be kept on ice for few hours.
3. Commercial polyP is often unavailable or discontinued. During the writing of the current chapter, Sigma-Aldrich offers polyP (sodium phosphate glass) Type 45 (45 phosphate groups as average polymeric length) (polyP45; S4379). In the past Sigma-Aldrich has offered polyP of variable polymeric length from 5 to 75.
4. Several commercially available IP_6 contain 5–10% of other inositol polyphosphates such as IP_5 and IP_7, which can be easily detected by toluidine staining. Therefore, it is advisable to check the purity of an IP_6 standard by PAGE before using it in important experiments. This is particularly relevant if the aim is to use IP_6 for enzymatic assays or binding experiments.
5. Commercial sources of toluidine blue differ in the amount of dye that they possess. Therefore it is important to assess the staining ability of each batch of toluidine blue.
6. Polyvinyl alcohol is difficult to dissolve. Heating and overnight stirring are often necessary. Filter the solution once fully dissolved.
7. Constructs to generate the recombinant proteins His-ScDdp1 and His-ScPpx1 are available from the Saiardi laboratory on request.
8. It is important not to overgrow *D. discoideum* liquid culture, since high cell density can induce differentiation. A detailed description of how to handle amoeba cells can be found on DictyBase (http://dictybase.org).
9. The pH of the mixed samples should be between 6 and 8. The pH can be measured by spotting a few μl of solution onto a pH

test strip, and can be adjusted accordingly using either perchloric acid or potassium carbonate with EDTA.

10. It is difficult to precisely normalize inositol polyphosphates by the amount of extracted cells/material. We can only rely on the initial number of cells that were plated to induce starvation. While cells do not divide in KK2, they aggregate to form slugs and fruiting bodies at the end of development, changing the amoeba physiology, cell shape, and cell number. Furthermore, it is complicated to collect the same percentage of biological material especially at the late stage of development, since *D. discoideum* cell architecture changes dramatically during development.
11. The described extraction copurifies polyP and RNA, therefore measuring the absorbance of nucleic acid at 260 nm allows normalization to the amount of extracted material.
12. The size of the glass plates and spacer in which the gel is prepared will determine the amount of gel solution to be prepared. To nicely resolve inositol polyphosphates we prepare 24 cm long, 1 mm thick polyacrylamide gels. For these gel dimensions we prepare 38 ml of gel solution. A complete description of the procedure used in our laboratory to prepare the polyacrylamide gels is available in video form [27].
13. Polyacrylamide polymerization is an exothermic process, and highly concentrated gels warm considerably during polymerization.
14. The correct voltage and amperage should be adjusted accordingly to the gel length and concentration used. The gel should be stopped when the Orange G dye arrives at 2/3 of the length of the gel. IP_6, IP_7, and IP_8 run faster than Orange G and may run out if the dye is allowed to run further than 2/3 of gel length.
15. The staining solution can be used multiple times until the dye is exhausted. We routinely prepare a fresh solution after 15–20 stainings.
16. Because of the polymeric nature of polyP, high polyacrylamide concentrations (e.g., 30%) can be used to properly resolve small polymeric polyP species, while 20% or lower polyacrylamide gel concentration is ideal to resolve large polymeric polyP species. Using different acrylamide concentrations dramatically affects gel migration thus voltage and amperage should be adjusted accordingly.
17. PolyP induced DAPI photobleaching [26] can take a couple of minutes to complete; therefore, it is opportune to record several images. After DAPI staining the gel can be subsequently stained with toluidine blue as described above.

Acknowledgments

We thank the Saiardi lab for suggestions and reading of the manuscript. This work was supported by the Medical Research Council (MRC) core support to the MRC/UCL Laboratory for Molecular Cell Biology University Unit MC_UU_12018/4 and MC_UU_00012/4. YD was supported by a Marie Skłodowska-Curie grant (agreement PHEMDD No 752903).

References

1. Azevedo C, Saiardi A (2014) Functions of inorganic polyphosphates in eukaryotic cells: a coat of many colours. Biochem Soc Trans 42 (1):98–102. https://doi.org/10.1042/BST20130111
2. Rao NN, Gomez-Garcia MR, Kornberg A (2009) Inorganic polyphosphate: essential for growth and survival. Annu Rev Biochem 78:605–647. https://doi.org/10.1146/annurev.biochem.77.083007.093039
3. Michell RH (2008) Inositol derivatives: evolution and functions. Nat Rev Mol Cell Biol 9 (2):151–161. nrm2334 [pii]. https://doi.org/10.1038/nrm2334
4. Livermore TM, Azevedo C, Kolozsvari B, Wilson MS, Saiardi A (2016) Phosphate, inositol and polyphosphates. Biochem Soc Trans 44 (1):253–259. https://doi.org/10.1042/BST20150215
5. Shears SB (2015) Inositol pyrophosphates: why so many phosphates? Adv Biol Regul 57:203–216. https://doi.org/10.1016/j.jbior.2014.09.015
6. Wilson MS, Livermore TM, Saiardi A (2013) Inositol pyrophosphates: between signalling and metabolism. Biochem J 452(3):369–379. https://doi.org/10.1042/BJ20130118
7. Lonetti A, Szijgyarto Z, Bosch D, Loss O, Azevedo C, Saiardi A (2011) Identification of an evolutionarily conserved family of inorganic polyphosphate endopolyphosphatases. J Biol Chem 286(37):31966–31974. M111.266320 [pii]. https://doi.org/10.1074/jbc.M111.266320
8. Wild R, Gerasimaite R, Jung JY, Truffault V, Pavlovic I, Schmidt A, Saiardi A, Jessen HJ, Poirier Y, Hothorn M, Mayer A (2016) Control of eukaryotic phosphate homeostasis by inositol polyphosphate sensor domains. Science 352(6288):986–990. https://doi.org/10.1126/science.aad9858
9. Odom AR, Stahlberg A, Wente SR, York JD (2000) A role for nuclear inositol 1,4,5-trisphosphate kinase in transcriptional control. Science 287(5460):2026–2029
10. Saiardi A, Caffrey JJ, Snyder SH, Shears SB (2000) The inositol hexakisphosphate kinase family. Catalytic flexibility and function in yeast vacuole biogenesis. J Biol Chem 275 (32):24686–24692
11. Saiardi A, Caffrey JJ, Snyder SH, Shears SB (2000) Inositol polyphosphate multikinase (ArgRIII) determines nuclear mRNA export in Saccharomyces cerevisiae. FEBS Lett 468 (1):28–32
12. Saiardi A, Erdjument-Bromage H, Snowman AM, Tempst P, Snyder SH (1999) Synthesis of diphosphoinositol pentakisphosphate by a newly identified family of higher inositol polyphosphate kinases. Curr Biol 9 (22):1323–1326
13. York JD, Odom AR, Murphy R, Ives EB, Wente SR (1999) A phospholipase C-dependent inositol polyphosphate kinase pathway required for efficient messenger RNA export. Science 285(5424):96–100
14. Saiardi A, Resnick AC, Snowman AM, Wendland B, Snyder SH (2005) Inositol pyrophosphates regulate cell death and telomere length through phosphoinositide 3-kinase-related protein kinases. Proc Natl Acad Sci U S A 102(6):1911–1914
15. Saiardi A, Sciambi C, McCaffery JM, Wendland B, Snyder SH (2002) Inositol pyrophosphates regulate endocytic trafficking. Proc Natl Acad Sci U S A 99(22):14206–14211
16. York SJ, Armbruster BN, Greenwell P, Petes TD, York JD (2005) Inositol diphosphate signaling regulates telomere length. J Biol Chem 280(6):4264–4269
17. Lander N, Ulrich PN, Docampo R (2013) Trypanosoma brucei vacuolar transporter chaperone 4 (TbVtc4) is an acidocalcisome

polyphosphate kinase required for in vivo infection. J Biol Chem 288(47):34205–34216. https://doi.org/10.1074/jbc.M113.518993

18. Zhang H, Gomez-Garcia MR, Shi X, Rao NN, Kornberg A (2007) Polyphosphate kinase 1, a conserved bacterial enzyme, in a eukaryote, dictyostelium discoideum, with a role in cytokinesis. Proc Natl Acad Sci U S A 104 (42):16486–16491. 0706847104 [pii]. https://doi.org/10.1073/pnas.0706847104
19. Livermore TM, Chubb JR, Saiardi A (2016) Developmental accumulation of inorganic polyphosphate affects germination and energetic metabolism in dictyostelium discoideum. Proc Natl Acad Sci U S A 113(4):996–1001. https://doi.org/10.1073/pnas.1519440113
20. Europe-Finner GN, Gammon B, Newell PC (1991) Accumulation of [3H]-inositol into inositol polyphosphates during development of dictyostelium. Biochem Biophys Res Commun 181(1):191–196
21. Newell PC, Europe-Finner GN, Small NV, Liu G (1988) Inositol phosphates, G-proteins and ras genes involved in chemotactic signal transduction of dictyostelium. J Cell Sci 89 (Pt 2):123–127
22. Stephens LR, Irvine RF (1990) Stepwise phosphorylation of myo-inositol leading to myo-inositol hexakisphosphate in Dictyostelium. Nature 346(6284):580–583
23. Stephens L, Radenberg T, Thiel U, Vogel G, Khoo KH, Dell A, Jackson TR, Hawkins PT, Mayr GW (1993) The detection, purification, structural characterization, and metabolism of diphosphoinositol pentakisphosphate(s) and bisdiphosphoinositol tetrakisphosphate(s). J Biol Chem 268(6):4009–4015
24. Losito O, Szijgyarto Z, Resnick AC, Saiardi A (2009) Inositol pyrophosphates and their unique metabolic complexity: analysis by gel electrophoresis. PLoS One 4(5):e5580. https://doi.org/10.1371/journal.pone.0005580
25. Pisani F, Livermore T, Rose G, Chubb JR, Gaspari M, Saiardi A (2014) Analysis of Dictyostelium discoideum inositol pyrophosphate metabolism by gel electrophoresis. PLoS One 9 (1):e85533. https://doi.org/10.1371/journal.pone.0085533
26. Smith SA, Morrissey JH (2007) Sensitive fluorescence detection of polyphosphate in polyacrylamide gels using 4′,6-diamidino-2-phenylindol. Electrophoresis 28 (19):3461–3465. https://doi.org/10.1002/elps.200700041
27. Loss O, Azevedo C, Szijgyarto Z, Bosch D, Saiardi A (2011) Preparation of quality inositol pyrophosphates. J Vis Exp (55):e3027. https://doi.org/10.3791/3027

Part II

Assays for Function

Chapter 6

Characterizing Enzymes of the Diphosphoinositol Polyphosphate Phosphohydrolase (DIPP) Family

Lucinda Winward, Rajagopal Sharada Kilari, and Stephen T. Safrany

Abstract

The diphosphoinositol polyphosphate phosphohydrolases are a subset of the Nudix hydrolase family of enzymes. As such, they metabolize a wide range of substrates, including diphosphoinositol polyphosphates (inositol diphosphates, inositol pyrophosphates), dinucleotide phosphates, nucleosides as well as 5-phosphoribosyl 1-pyrophosphate and inorganic polyphosphate. Here, we describe protocols to optimize these enzymes, with consideration to buffer composition and sample preparation and how to analyze the metabolism of these substrates using high-performance liquid chromatography, giving advice where pitfalls are commonly encountered.

Key words Inositol diphosphate, Inositol pyrophosphate, Nudix, DIPP, Diadenosine polyphosphate, Polyphosphate

1 Introduction

The most phosphorylated members among the inositol phosphates are the diphosphoinositol polyphosphates (DIPs), also known as inositol diphosphates or inositol pyrophosphates (for reviews *see* refs. 1, 2). These are phosphorylation products of the two most abundant inositol phosphates found in cells, inositol 1,3,4,5,6-pentakisphosphate (InsP_5) and inositol hexakisphosphate (InsP_6) [3]. Two diphosphoinositol pentakisphosphates (PP-InsP_5) have been identified as precursors to bisdiphosphoinositol tetrakisphosphate ([PP]$_2$-InsP_4), as well as a diphosphoinositol tetrakisphosphate (PP-InsP_4). The enzymes involved in the hydrolysis of these DIPs are members of the Nudix hydrolase superfamily. These enzymes predominantly cleave the diphosphate linkage in compounds of general structure NDP–X (nucleoside diphosphate linked to another moiety X) with varying degrees of specificity, although some have activity toward nonnucleotide pyrophosphates, such as DIPs. Diphosphoinositol polyphosphate phosphohydrolases (DIPPs) hydrolyze DIPs but most also have activity

Gregory J. Miller (ed.), *Inositol Phosphates: Methods and Protocols*, Methods in Molecular Biology, vol. 2091,
https://doi.org/10.1007/978-1-0716-0167-9_6,

toward diadenosine 5′,5‴-P^1,P^5-pentaphosphate (Ap_5A) and diadenosine 5′,5‴-P^1,P^6-hexaphosphate (Ap_6A). The human enzymes termed hDIPP-1 (NUDT3) [4, 5], hDIPP-2 (isoforms α and β) (NUDT4) [6], hAps1 (hDIPP-3β or NUDT11) and hAps2 (hDIPP3α or NUDT10) [7, 8] as well as *Drosophila melanogaster* Aps [9], *Schizosaccharomyces pombe* Aps1 [10], *Saccharomyces cerevisiae* Ddp1p [11] and AtNUDT13 from *Arabidopsis thaliana* [12] all favor Ap_6A over Ap_5A. Activity with diadenosine 5′,5‴-P^1,P^4-tetraphosphate (Ap_4A) is lower, and is in fact lacking in the case of Ddp1p and AtNUDT13. PP-InsP_5 and $[PP]_2$-InsP_4 are generally favored in vitro, with k_{cat}/K_m ratios being 4- to 500-fold higher for PP-InsP_5 compared to Ap_6A. A poly-His-tagged fusion protein of hDIPP-2β was originally found to have reduced activity toward DIPs compared to nucleotide substrates [6]. Likewise, the poly-His-tagged AtNUDT13 from *Arabidopsis thaliana* been reported to lack PP-InsP_5 hydrolase activity altogether [12]. Subsequently hDIPP-2β was found to have activity similar to the other human DIPPs when expressed as a GST-fusion protein [13]. It has also been shown that DIPP family members have 5-phosphoribosyl 1-pyrophosphate pyrophosphatase activity that generates the glycolytic activator ribose 1,5-bisphosphate [14], and inorganic polyphosphate [15].

2 Materials

2.1 Preparation of PP-Ins

Buffer A: 1 mM disodium ethylene diamine tetraacetic acid (Na_2EDTA), 6 mM $MgSO_4$, 5 mM ATP, 1 mM dithiothreitol, 20 mM HEPES (pH 6.8), 4 mM CHAPS.

4.6 × 125 mm Partisphere 5 μm SAX HPLC column.

50% (v/v) ammonia.

Sep-Pak Accell Plus QMA cartridge (Waters).

Triethylammonium bicarbonate (TEAB), 1.5 M, pH 7.

2.2 Analysis of Inositol Phosphate Metabolism

Buffer B: 50 mM KCl, 50 mM HEPES (pH 7.2 with KOH), 4 mM CHAPS, 50 μg/ml bovine serum albumin (BSA), 1 mM Na_2EDTA (*see* **Note 1**), 2 mM $MgSO_4$.

8% perchloric acid.

Potassium carbonate (2 M K_2CO_3, 5 mM Na_2EDTA).

Universal indicator.

Buffer C: 10 mM $MgCl_2$, 50 μg/ml BSA and 15 mM MES, 15 mM HEPES and 15 mM Tris (pH 6–9 with HCl or NaOH).

Buffer D: 50 mM KCl, 50 mM HEPES (pH 7.2 with KOH), 4 mM CHAPS, 50 μg/ml BSA, 1 mM Na_2EDTA, and 50 μM–10 mM (where possible) chloride salts of divalent cations.

2.3 Analysis of Nucleotide Metabolism

Buffer E: 50 mM HEPES (pH 7.6 with KOH), 1 mM $MnCl_2$ and 50 μg/ml BSA.

Manganese (II) chloride ($MnCl_2$).

2.4 Separation of PP-InsPs

4.6 × 125 mm Partisphere 5 μm SAX HPLC column.

Pro Flow-P+ scintillation cocktail.

Buffer F: 1 mM Na_2EDTA, no pH adjustment.

Buffer G: buffer A plus 1.3 M $(NH_4)_2HPO_4$, pH 3.85 with H_3PO_4.

Buffer H: 1.3 M $(NH_4)_2HPO_4$, pH 4.8 with H_3PO_4.

2.5 Analysis of Nonradioactive PP-Ins

Toluidine blue.

A wide range of nucleotides and analogues are available from Jena Bioscience, Jena, Germany (www.jenabioscience.com). Other nucleotides are available from Sigma (www.sigmaaldrich.com) and TriLink Biotechnologies (www.trilinkbiotech.com). Further nucleotides and analogues are available through scientific collaboration with Prof. Dr. Henning J. Jessen (www.jessen-lab.uni-freiburg.de/, e-mail: henning.jessen@oc.uni-freiburg.de). Nonradioactive PP-InsP_5 (several isomers) and $[PP]_2$-InsP_4 are, again, available from Prof. Dr. Jessen. [^{3}H] InsP_6 is available from American Radiolabeled Chemicals, (http://www.arcincusa.com/) and can be found under "Phytic Acid."

3 Methods

3.1 Preparation of Reagents

5-PP-[^{3}H] InsP_5, 1-PP-[^{3}H] InsP_5 and 1,5 $[PP]_2$-[^{3}H] InsP_4 can be prepared from [^{3}H] InsP_6 using recombinant Kcs1p and Vip1p [16]. Briefly, [^{3}H] InsP_6 is incubated with either Kcs1p, Vip1p or both in Buffer A at 30 °C until >90% of conversion has been achieved. Preparation of [^{32}P] or [^{33}P] reagents can also be achieved using unlabeled InsP_5 or InsP_6 and [γ-^{32}P] or [γ-^{33}P] ATP. The low affinity of Kcs1p and Vip1p for ATP encourages the user to prepare reagents of lower specific activity, however, reducing the concentration of unlabeled ATP will enhance the specific activity of the product. Unless the ultimate destination of the phosphate(s) added are of interest, as with protein pyrophosphorylation assays [17], we would advise the use of [^{3}H]-labeled materials. Should purification be required (as it will in all cases using [γ-^{32}P] or [γ-^{33}P] ATP), high performance liquid chromatography (HPLC) using a Partisphere SAX (HiChrom) column (see below) followed by desalting is advised. Following HPLC, samples should be neutralized with an appropriate volume of 50% (v/v) ammonia, diluted 20-fold, and then loaded onto a Sep-Pak Accell Plus QMA cartridge. Columns are washed with 8 ml of 0.4 M triethylammonium bicarbonate (TEAB) (pH 7 with CO_2 (*see* **Note 2**)) and then products eluted with 5 ml of 1.5 M TEAB (pH 7), before drying overnight using a centrifugal evaporator (*see* **Note 3**).

3.2 Analysis of Inositol Phosphate Metabolism

The temperature at which these experiments should be undertaken depends on the body temperature of the organism it has been taken from, although some proteins do show unexpected temperature profiles. Reactions are quenched by the addition of an equal volume of ice-cold 8% perchloric acid. This causes protein precipitation. Following incubation on ice (20 min, 4 °C), this protein must be removed. Centrifugation and removal of the supernatant is imperative, as not all DIPP-like enzymes are permanently inactivated by acid precipitation (*see* **Note 4**). The supernatant is then neutralized by the addition of potassium carbonate (2 M K_2CO_3, 5 mM Na_2EDTA). The addition of universal indicator (10 μl) prior to this step facilitates in attaining a neutral sample (straw colored). Excess K_2CO_3 can cause the material to become alkaline and damage the silica-based HPLC column. Addition of K_2CO_3 permits the formation of potassium perchlorate precipitate and release of gaseous carbon dioxide (*see* **Note 5**).

Should the pH optimum of an enzyme be of interest, we advocate the use of a single buffer (Buffer C) comprising 10 mM $MgCl_2$, 50 μg/ml BSA and 15 mM MES, 15 mM HEPES and 15 mM Tris (pH 6–9 with HCl or NaOH) rather than preparing a series of different buffers in which different pH buffers are used in isolation. This prevents particular components inadvertently affecting enzyme activity in one particular pH range, as can be seen elsewhere [18].

Divalent ion-dependency can be determined in Buffer D comprising 50 mM KCl, 50 mM HEPES (pH 7.2 with KOH), 4 mM CHAPS, 50 μg/ml BSA, 1 mM Na_2EDTA, and 50 μM–10 mM (where possible) divalent metal as the chloride salt (*see* **Note 6**). The pH adjustment should be performed before addition of a concentrated solution of divalent cation (*see* **Note 7**). It is noted that some metals form insoluble complexes with inositol phosphates and will not yield results. It is the purpose of the Na_2EDTA to chelate out low levels of such contaminating metals in other salts.

One DIPP (Aps from *D. melanogaster*) has been tested for the effects of reducing agent: it was observed that under reducing conditions Aps was thermostable, unaffected by a 5 min treatment at 76 °C. In comparison, oxidized Aps was completely inactivated by a 5 min treatment at 56 °C. The dimerization apparent was not observed previously with DIPP-1, but sensitivity to heat was not tested. The only other DIPP with a CIC motif near the N-terminal, Aps1 from *S. cerevisiae*, was also found to have "an absolute requirement for reducing agents" [19]. A second, non-Nudix enzyme, from *S. cerevisiae* (Siw14) [20] may also be subject to redox regulation due to a catalytic cysteine resembling PTEN [21].

3.3 Analysis of Nucleotide Metabolism

For the analysis of nucleotide metabolism, a common Buffer E can be used. Unless specifically required, Na_2EDTA should not be added (*see* **Note 8**). Divalent ion-dependency can be determined in the same buffer, exchanging 1 mM $MnCl_2$ for chloride salts of other divalent metals at 50 μM–10 mM (where possible).

3.4 HPLC Analysis of Rates and Routes of Hydrolysis

Materials can be separated using a 4.6 × 125 mm Partisphere 5 μm SAX HPLC column. We have found no improvement by using a 4.6 × 250 mm column. Using a flow rate of 1 ml/min, we recommend the following gradients: the optimal gradient for inositol phosphates utilises Buffer F (1 mM Na_2EDTA, no pH adjustment) and Buffer G (buffer A plus 1.3 M $(NH_4)_2HPO_4$, pH 3.85 with H_3PO_4 (*see* **Note 9**) created by mixing Buffer F and Buffer G as follows: 0–5 min, 0% Buffer G; 5–10 min, 0–50% Buffer G; 10–60 min, 50–100% Buffer G; 60–70 min, 100% Buffer G; 71–91 min, 0% Buffer G. Fractions are collected at 1-min intervals, beginning 15 min into the gradient. We advise the use of Pro Flow-P+ scintillation cocktail modified for high phosphate concentrations by Meridian Biotech (www.meridian-biotech.com/) at a ratio of 1:3 (sample–cocktail) when assessing radioactivity. We have not found any commercial, off-the-shelf scintillant to work as effectively as this product. Indeed, unless specified as being suitable for high phosphate, low pH samples, counting efficiency will be essentially zero.

The separation of nucleosides requires a different pH (at pH 3.8 some products of hydrolysis coelute with substrates). Therefore, we recommend the use of a gradient created by mixing water with Buffer H (*see* **Notes 9** and **10**): 0–5 min, 0% buffer H; 5–55 min, 0–50% buffer H; 56–70 min, 0% Buffer H. Under these conditions all substrates can be separated from their subsequent products.

3.5 Analysis of Nonradioactive Inositol Phosphates

The methods described above only permit the analysis of radioactive materials of biologically prepared materials (1-PP-InsP_5, 5-PP-InsP_5, and $[PP]_2$-InsP_4). Nonradioactive analyses have been performed on these and other synthetic substrates. We would advise caution, as the limit of detection (2 nmol with toluidine blue [22]) allows experiments to be performed in the μM range [20], whereas the K_m for most enzymes determined using radiolabeled substrates fall in the nM range [13]. This would suggest that accurate affinity measurements are well outside the range of detection of these protocols at the moment. We rely on researchers in the field to continue pursuing such endeavors and trust that further improvements will be realized. Such technology already exists: the metal dye detection system developed by Georg Mayr shows much greater sensitivity [23]. The technical difficulties and specialist requirements of such a system have meant it has not been widely adopted.

4 Notes

1. Na_2EDTA (1 mM) should be added to all buffers when analysing inositol phosphates to maintain solubility of higher inositol phosphates.
2. This can be achieved by bubbling water-saturated CO_2 through the TEAB buffer. Caution should be taken, as pH can change quickly as the desired pH approaches.
3. The HPLC and subsequent desalting leads to significant losses. We suggest optimization of the kinase step, which would negate the need for this purification and desalt, as the carry-over of salts and ATP is minimal if specific activity of the product is sufficiently high.
4. Aps was found to retain full activity after acid precipitation and recovery by neutralization.
5. Addition of K_2CO_3 should be performed drop-wise with mixing. The rapid production of CO_2 can cause samples to bubble over, causing loss of sample and potential radioactive contamination of the work area.
6. Free concentration of divalent cations is calculated using Max-Chelator (http://www.stanford.edu/~cpatton/maxc.html). It is the purpose of the Na_2EDTA to chelate out low levels of contaminating metals present in the salts used.
7. The addition of hydroxide to buffers containing some divalent metals will cause precipitation.
8. Na_2EDTA binds the HPLC column and, when eluted, absorbs at 254 nm, masking some of the substrates or products.
9. The amount required frequently surprises novice researchers: this requires approximately 90 ml of concentrated H_3PO_4/L buffer to obtain pH 3.85 or 65 ml of concentrated H_3PO_4/L buffer to obtain pH 4.8.
10. The quality of H_3PO_4 should be at least ACS grade (we use Sigma-Aldrich 438081), as cheaper products absorb in the UV range and interfere with data acquisition as well as increase back pressure of the system.

References

1. Shears SB (2016) Towards pharmacological intervention in inositol pyrophosphate signalling. Biochem Soc Trans 44:191–196
2. Padmanabhan U, Kilari RS, Winward LJ et al (2016) Chapter 5 Inositol diphosphates: an expanding repertoire of functions and regulation. In: Phytate destruction – consequences for precision animal nutrition. Wageningen Academic Publishers, Wageningen, pp 61–86
3. Safrany ST, Caffrey JJ, Yang X et al (1999) Diphosphoinositol polyphosphates: the final frontier for inositide research? Biol Chem 380 (7–8):945–951
4. Safrany ST, Caffrey JJ, Yang X et al (1998) A novel context for the 'MutT' module, a guardian of cell integrity, in a diphosphoinositol polyphosphate phosphohydrolase. EMBO J 17(22):6599–6607
5. Safrany ST, Ingram SW, Cartwright JL et al (1999) The diadenosine hexaphosphate hydrolases from *Schizosaccharomyces pombe* and Saccharomyces cerevisiae are homologues of the human diphosphoinositol polyphosphate phosphohydrolase. Overlapping substrate specificities in a MutT-type protein. J Biol Chem 274(31):21735–21740
6. Caffrey JJ, Safrany ST, Yang X et al (2000) Discovery of molecular and catalytic diversity among human diphosphoinositol-polyphosphate phosphohydrolases. An expanding Nudt family. J Biol Chem 275 (17):12730–12736
7. Leslie NR, McLennan AG, Safrany ST (2002) Cloning and characterisation of hAps1 and hAps2, human diadenosine polyphosphate-metabolising Nudix hydrolases. BMC Biochem 3(1):20
8. Hidaka K, Caffrey JJ, Hua L et al (2002) An adjacent pair of human NUDT genes on chromosome X are preferentially expressed in testis and encode two new isoforms of diphosphoinositol polyphosphate phosphohydrolase. J Biol Chem 277(36):32730–32738
9. Winward L, Whitfield WG, McLennan AG et al (2010) Oxidation of the diphosphoinositol polyphosphate phosphohydrolase-like Nudix hydrolase Aps from *Drosophila melanogaster* induces thermolability--A possible regulatory switch? Int J Biochem Cell Biol 42 (7):1174–1181
10. Ingram SW, Stratemann SA, Barnes LD (1999) *Schizosaccharomyces pombe* Aps1, a diadenosine 5′,5′″-P^1, P^6- hexaphosphate hydrolase that is a member of the nudix (MutT) family of hydrolases: cloning of the gene and characterization of the purified enzyme. Biochemistry 38 (12):3649–3655
11. Cartwright JL, McLennan AG (1999) The *Saccharomyces cerevisiae* YOR163w gene encodes a diadenosine 5′, 5‴- P^1,P^6-hexaphosphate (Ap_6A) hydrolase member of the MutT motif (Nudix hydrolase) family. J Biol Chem 274 (13):8604–8610
12. Olejnik K, Murcha MW, Whelan J et al (2007) Cloning and characterization of AtNUDT13, a novel mitochondrial *Arabidopsis thaliana* Nudix hydrolase specific for long-chain diadenosine polyphosphates. FEBS J 274 (18):4877–4885
13. Kilari RS, Weaver JD, Shears SB et al (2013) Understanding inositol pyrophosphate metabolism and function: kinetic characterization of the DIPPs. FEBS Lett 587(21):3464–3470
14. Fisher DI, Safrany ST, Strike P et al (2002) Nudix hydrolases that degrade dinucleoside and diphosphoinositol polyphosphates also have 5-Phosphoribosyl 1-Pyrophosphate (PRPP) pyrophosphatase activity that generates the glycolytic activator ribose 1,5-Bisphosphate. J Biol Chem 277 (49):47313–47317
15. Lonetti A, Szijgyarto Z, Bosch D et al (2011) Identification of an evolutionarily conserved family of inorganic polyphosphate endopolyphosphatases. J Biol Chem 286 (37):31966–31974
16. Loss O, Azevedo C, Szijgyarto Z et al (2011) Preparation of quality inositol pyrophosphates. Jove 2011(55):e3027
17. Saiardi A, Bhandari R, Resnick AC et al (2004) Phosphorylation of proteins by inositol pyrophosphates. Science 306(5704):2101–2105
18. Liew SM, Tay ST, Puthucheary SD (2013) Enzymatic and molecular characterisation of leucine aminopeptidase of *Burkholderia pseudomallei*. BMC Microbiol 13:110
19. Cartwright JL, McLennan AG (1997) The MutT motif-containing ORF163w protein from the yeast *Saccharomyces cerevisiae* encodes a (di)nucleoside polyphosphate hydrolase. Biochem Soc Trans 25(4):S580
20. Steidle EA, Chong LS, Wu M et al (2016) A novel inositol pyrophosphate phosphatase in *Saccharomyces cerevisiae*: Siw14 protein

selectively cleaves the beta-phosphate from 5-diphosphoinositol pentakisphosphate (5PP-IP_5). J Biol Chem 291(13):6772–6783

21. Ross SH, Lindsay Y, Safrany ST et al (2007) Differential redox regulation within the PTP superfamily. Cellular Signal 19(7):1521–1530
22. Losito O, Szijgyarto Z, Resnick AC et al (2009) Inositol pyrophosphates and their unique metabolic complexity: analysis by gel electrophoresis. PLoS One 4(5):e5580
23. Mayr GW (1988) A novel metal-dye detection system permits picomolar-range HPLC analysis of inositol polyphosphates from non-radioactively labeled cell or tissue specimens. Biochem J 254(2):585–591

Chapter 7

Metabolic Labeling of Inositol Phosphates and Phosphatidylinositols in Yeast and Mammalian Cells

Andrew T. Hale, Bradley P. Clarke, and John D. York

Abstract

Inositol phosphate (IP) and phosphatidylinositol (PI) signaling are critical signal transduction pathways responsible for generating numerous receptor-mediated cellular responses. Biochemical and genetic studies have revealed diverse roles of IP and PI signaling in eukaryotic signaling, but detailed characterization of unique IP and PI signaling profiles in response to different agonists and among cell types remains largely unexplored. Here, we outline steady-state inositol metabolic-labeling techniques that can be leveraged to assess the IP and PI signaling state in eukaryotic cells. This flexible technique can be amended and optimized to your cell line of interest, perturbed with biochemical, genetic, or pharmacological alteration, and used to provide comprehensive inositol profiling in various cellular systems.

Key words Inositol phosphates, Phosphatidylinositol, Metabolic labeling, Signal transduction, HPLC, Lipid signaling

1 Introduction

Phosphorylation is a ubiquitous signal used by cells to control biological processes. In addition to the widely studied phenomenon of protein phosphorylation, small molecules can also serve as cellular messengers. One such molecule is *myo*-inositol, a cyclohexanehexol obtained primarily from dietary sources [1–3]. In eukaryotic cells, soluble inositol phosphates (IPs) are derived from the lipid carrier phosphatidylinositol-4,5-bisphosphate (PIP_2). Stimulation of phospholipase C (PLC) cleaves PIP_2 to form diacylglycerol and inositol-1,4,5-trisphosphate (IP_3), a classic second messenger that stimulates Ca^{2+} release from the sarcoplasmic reticulum [1, 2]. IP_3 can be phosphorylated at all six carbons and diphosphorylated at two, leading to over a dozen chemically unique molecules with distinct roles in regulating transcription, mRNA export, telomere maintenance, embryonic development, cell division, nutrient sensing, and acting as structural cofactors [4–10]. Inositol signaling is complex, with various kinases and

Gregory J. Miller (ed.), *Inositol Phosphates: Methods and Protocols*, Methods in Molecular Biology, vol. 2091,
https://doi.org/10.1007/978-1-0716-0167-9_7, © Springer Science+Business Media, LLC, part of Springer Nature 2020

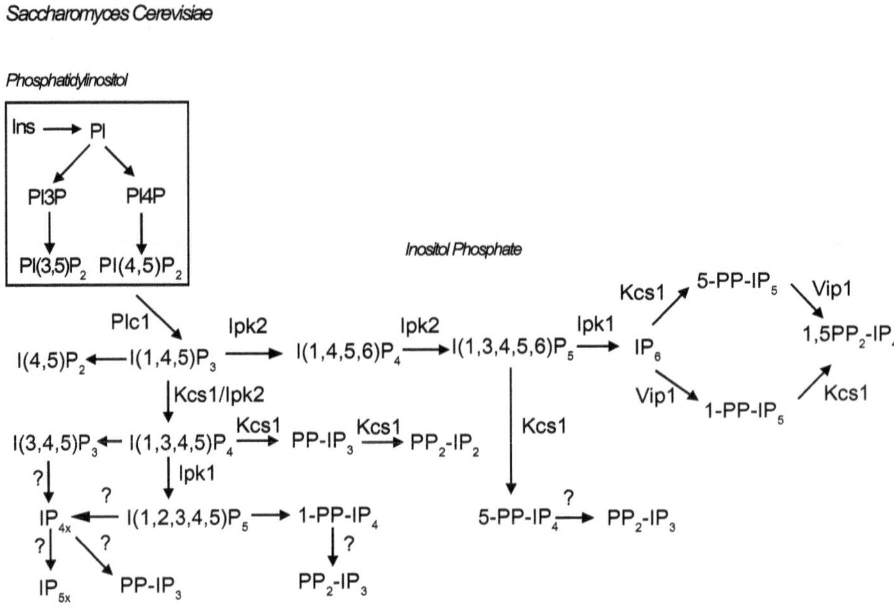

Fig. 1 Diagram of the inositol phosphate and phosphatidylinositol signaling pathways in *Saccharomyces cerevisiae*. (Adapted from ref. 29)

phosphatases acting on IPs to create an array of small molecules amongst eukaryotic species (*Saccharomyces Cerevisiae*—Fig. 1; *Dictyostelium*—Fig. 2; Metazoans—Fig. 3). Therefore, equally diverse IP profiles are expected given the heterogeneity of cellular functions mediated through IP signaling.

Similarly complex, PIPs can be phosphorylated or dephosphorylated and act as lipid signaling molecules capable of coordinating cell growth, division and acting as direct modifiers of nuclear and cell-surface receptors [11–17], in addition to a number of other cell signaling actions. Although PI comprises a minority of total cell-membrane lipid, its role in signaling is critical. Free PI generates PI(3)P, PI(4)P, and PI(5)P through a number of kinases; PI(X)Ps can be further phosphorylated to both PIP_2 and PIP_3 species (Fig. 3). Genetic disruption and dysregulation of these numerous kinases and phosphatases have been implicated in a number of congenital and acquired diseases [18–23]. Similarly, PIPs can also be independently activated by PI3-K, which triggers Akt-induced activation of cellular growth, movement, and proliferation [12, 16], among other responses. Alteration of PI3-K signaling is observed in a number of diseases as well [17, 20]. Thus, interrogation of the specific IP and/or PIP profiles in response to pharmacologic, chemical or genetic manipulation offers precise molecular signaling details in the biological system being tested.

Here, we detail experimental procedures to measure IPs and PIPs via steady-state metabolic labeling in yeast and mammalian cells. This technique can be modified to measure IPs and PIPs in

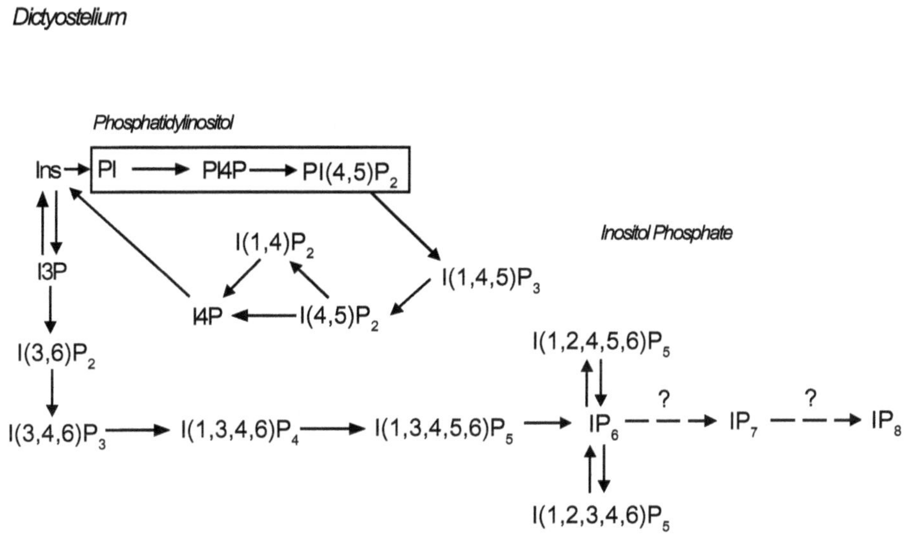

Fig. 2 Diagram of the inositol phosphate and phosphatidylinositol signaling pathways in *Dictyostelium*. (Adapted from ref. 31)

numerous cell types, including postmitotic cells. We anticipate that the techniques described herein enable investigators to discover new roles for IPs and PIPs in various eukaryotic cell systems.

2 Materials

2.1 Cell Culture

1. Cell line of interest (HEK293T, Cos7, Rat1, Hela, HT29, MEFs, etc.).
2. 6 or 12-well culture dishes (Corning, NY).
3. Dulbecco's Modified Eagle Medium (DMEM) (Sigma-Aldrich, St. Louis, MO) or Roswell Park Memorial Institute (RPMI) (Thermo Fisher, Cambridge, MA) media supplemented with 10% fetal bovine serum (Thermo Fisher, Cambridge, MA) and 1X penicillin/streptomycin (Sigma-Aldrich, St. Louis, MO).
4. Inositol-free Dulbecco's Modified Eagle Medium (DMEM) (Sigma-Aldrich, St. Louis, MO) or inositol-free Roswell Park Memorial Institute (RPMI) (Thermo Fisher, Cambridge, MA) media supplemented with 10% dialyzed fetal bovine serum Thermo Fisher, Cambridge, MA) and 1× penicillin/streptomycin (Sigma-Aldrich, St. Louis, MO). We anticipate other media types can be used as long as they can be depleted of inositol.
5. 1× Phosphate buffered saline (PBS) (Sigma-Aldrich, St. Louis, MO).

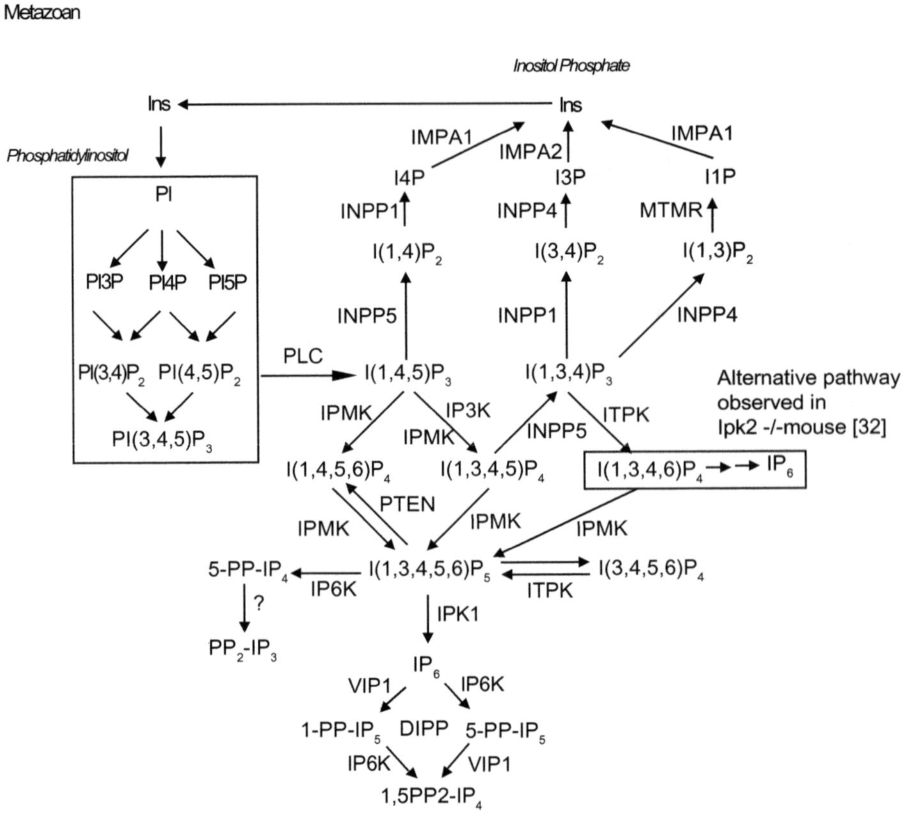

Fig. 3 Diagram of the inositol phosphate and phosphatidylinositol signaling pathways in metazoans. To determine if a novel pyrophosphate is derived from IP_5 or IP_6, one can treat the fraction collected by HPLC with diphosphoinositol phosphate phosphatase (DIPP) as previously described [26]. Identification of the alternative pathway to generate IP_6 is detailed previously [32]

6. Hydrochloric acid (diluted to 0.5 M) (Sigma-Aldrich, St. Louis, MO).
7. *Myo*-^{3}H-inositol (Perkin-Elmer, Waltham MA) at 37.5–200 μCi/mL (depending on length of labeling experiment and cell type of interest). Alternatively, C^{14} and P^{32} can be used. However, the methods outlined here pertain specifically to ^{3}H-inositol.

2.2 High Performance Liquid Chromatography (HPLC)

1. 1.7 M ammonium phosphate (Fisher Scientific, Fairlawn NJ) at a pH of 3.5. 10 mM ammonium phosphate is prepared by dilution with deionized water and passed through a 0.22 μM cellulose acetate bottle top sterile filter (Corning Inc., Corning, NY) prior to use in autoclave-sterilized buffer containers.
2. Buffer A: 10 mM ammonium phosphate (pH = 3.5). Buffer B: 1.7 M ammonium phosphate (pH = 3.5).

3. The HPLC system (Waters Bechmann or Agilent) equipped with an autosampler (desired) should contain a 4.6 × 125 mm Partisphere SAX strong anion exchange column (Whatman, Maidstone, England). Scintillation fluid (Perkin Elmer, Fairlawn, NJ, Catalog number 50-905-0584) is collected at a ratio of 6 mL scintillation fluid per 1 mL HPLC fraction in 7 mL scintillation vials (Fisher Scientific, Fairlawn, NJ) [24].
4. Scintillation counts are performed using a Tri-Carb 2100 TR (Packard, Fairlawn, NJ).
5. It is also possible (desired) to have an in-line radioactivity detection system such as INUS beta-ram detector [24].

2.3 Soluble IP and PIP Extraction

1. 372 μL $CHCl_3$ and MeOH (Fisher Scientific, Fairlawn, NJ) at a ratio of 1:2 is added to 100 μL of 0.5 M HCl extracted cells. In yeast, acid-washed bead beating (OPS Diagnostics, Lebanon, NJ) is required to lyse the thick cell wall.
2. 100 mg/mL bovine brain extract (Sigma-Aldrich, St. Louis, MO) in $CHCl_3$ (Fisher Scientific, Fairlawn, NJ).
3. For deacylation and PIP extraction, a speed vacuum (Thermo Fisher, Cambridge, MA), propanol, 33% methylamine, butanol, petroleum ether, ethyl formate, and glass tubes (Fisher Scientific, Fairlawn, NJ) will be needed.

3 Methods

Interrogation of IP and PIPs in steady-state equilibrium enables robust analysis of genetic, biochemical or pharmacological manipulations of inositol signaling. Our group has published extensively in IP labeling in a variety of cell types and organisms [8, 24–30]. Here, we offer troubleshooting tips and guidelines for metabolic labeling in varying cell lines. The protocol described below details measurement of IPs and PIPs in yeast and mammalian cells by HPLC.

3.1 Cell Culture and Isolation of IPs and PIPs

1. Maintain cell line in regular growth media, then exchange inositol-complete to inositol-free medium for at least 24 h prior to radiolabeling.
2. Plate your cell line of interest approximately 100,000–250,000 cells in 1 mL media (density of ~10,000–30,000 cells/cm^2 in a 6-well culture dish and allow cells to equilibrate overnight. Cell lines that are both adherent to the dish and in suspension can be used.
3. Supplement inositol-free media with 37.5–200 μCi/mL of radiolabeled inositol and allow cells to double approximately 2–4 times. Depending on the cell line, this may range from 2 to 7 (or more) days.

4. Once cell line is near confluence, remove media, wash with 1× PBS and add 0.5 M HCl for 5 min prior to manual scraping of cells from surface of plate. Amount of HCl can be varied according to dish size and cell count. Alternatively, cells can be scraped via manual force in PBS or trypsinized, and HCl can be added directly to the cell pellet.
5. For yeast, grow overnight culture of yeast in complete synthetic medium. Inoculate 1 mL of fresh complete synthetic medium containing 20–200 μCi/mL ^{3}H-inositol with 10 μL of overnight culture and grown to log phase or saturation.
6. Harvest yeast by centrifugation (5 min at 16,873 × *g*). Aspirate old medium and wash cells with 1 mL of PBS. Spin down cells, aspirate PBS and resuspend cell pellet in 100 μL of 0.5 M HCl.

3.2 Extraction of Soluble IPs and PIPs

1. Per 100 μL of 0.5 M HCl, add 372 μL of $CHCl_3$–MeOH (1 part $CHCl_3$ and 2 parts MeOH) and 100 μL acid-washed beads (for yeast) or vortex mammalian cells. Then, add 125 μL $CHCl_3$ and 125 μL 2 M KCl (do not make master mix), followed by an additional 2 min of bead beating. For mammalian cells, vortex for approximately 15 s or until thoroughly homogeneous. Next, spin down for 5 min at 16,873 × *g*.
2. Carefully separate the upper aqueous and lower organic layers. The upper aqueous layer contains soluble IPs, whereas the lower organic layer contains phosphatidylinositols.
3. For PIP extraction, add 10 μL of 100 mg/mL of bovine brain extract in $CHCl_3$ and dry speed vacuum at 35 °C for approximately 1 h or until sample is dry.
4. Next, add 300 μL of deacylation reagent (10 parts 33% methylamine in ethanol, 3 parts water, and 1 part butanol; mix in glass tube; scale master mix accordingly) followed by 2 h of shaking at 53 °C. Perform all handling of methylamine in a chemical hood as this reagent is highly flammable. Once complete, spin down to pellet, add 150 μL of ice-cold propanol, and dry speed vacuum at 80 °C. Drying time will vary depending on the volume in your sample.
5. Once the speed vacuum cycle is complete, dissolve in 200 μL water and 300 μL extraction solution (20 parts 1-butanol, 4 parts petroleum ether, 1 part ethyl formate; mix in glass tube and keep on ice); vortex to mix. Spin down for 5 min at 14,000 rpm and retain lower phase.
6. Add an additional 300 μL of extraction solution, vortex, spin down for 5 min at 14,000 rpm and retain lower phase. Repeat this **step 2** additional times, retaining the lower phase. Dilute this phase as necessary in 10 mM ammonium phosphate and sterile filter prior to HPLC analysis.

3.3 High Performance Liquid Chromatography

100 μL of sample is injected onto a 1 mL Partisphere SAX column. Flow rate is maintained at 1 mL per minute and 1 mL fractions are manually collected into 7 mL scintillation vials (Tables 1 and 2). Once the run is complete, add 6 mL of scintillation fluid to each vial and count use a scintillation counter. Example HPLC traces for IP and PIP analysis can be found in Fig. 4.

Table 1
HPLC gradient for IP analysis

Time (min)	% Buffer B	Approximate retention time (min)
0–5	0	
5–70	Linear gradient to 100%	IP_3 (33) IP_4 (45) IP_5 (58)
70–100	100%	IP_6 (75)

Table 2
HPLC gradient for PIP analysis

Time (min)	% Buffer B	Approximate retention time (min)
0–5	0	
5–12	0–7	PI (5)
12–15	7	
15–30	7–14	
30–40	14–60	PI(3)P (15) PI(4)P (17) PI(3,5)P_2 (36) PI(4,5)P_2 (38)
40–45	60	PI(3,4,5)P_3 (42)
45–55	60–100	
55–65	100	
65–80	0	

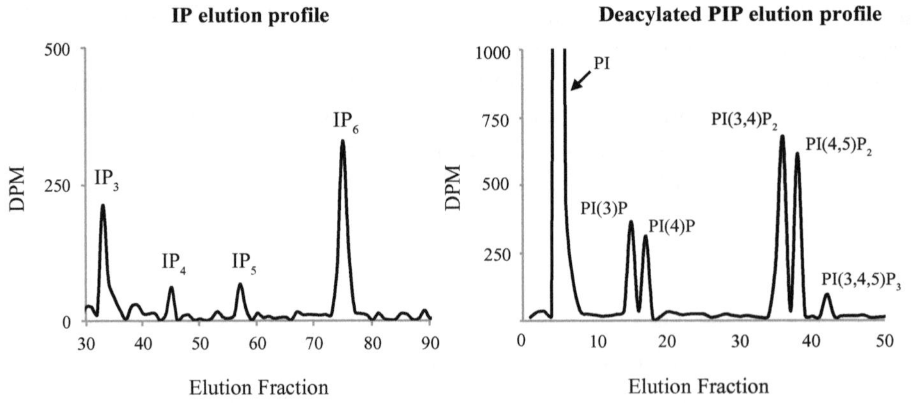

Fig. 4 Example IP HPLC trace from HT29 cells and PIPs from T lymphocytes. Elution times can vary depending on the column. Thus, we recommend using radiolabeled yeast standards when initiating a new experiment or changing columns

4 Notes

1. The total number of cell doublings, rather than the time cells are maintain in radiolabeled inositol determines the degree of saturation toward steady state and strength of scintillation counts detected. However, postmitotic cells can be efficiently labeled by increasing the concentration of radiolabeled inositol and prolonged the labeling time in vitro.
2. Similarly, the total concentration of radiolabeled inositol used in culture varies from cell line to cell line, required optimization by the investigator. The concentration of radiolabeled inositol and labeling length of time should be empirically determined for each cell line of interest.
3. Typically, collection of 1 mL fractions every minute during the HPLC gives the highest resolution of IP and PI species.
4. Although automated beta-ram detectors have been employed to measure inositol species, we find that manual scintillation counts are often more sensitive to detection of IPs.
5. A comprehensive set of standards should be run for accurate determination of IP and PI species identification as elution times can vary depending on the column.
6. As described previously [28], higher-order IPs (IP_7, IP_8) may require transduction of inositol phosphate multikinase (IPMK) or robust activation of G protein-signaling via transfection of a constitutively active G protein-coupled receptor. In addition, to identify inositol pyrophosphates derived from IP_5 or IP_6, elution fractions can be treated with diphosphoinositol phosphate phosphatase (DIPP) as previously described [26].

7. Due to sample-to-sample variation, raw counts obtained by scintillation should be interpreted with caution as saturation of inositol labeling is not always 100%. Normalization to free inositol or IP_3 can be useful, but interpretation of these results should be done with rigor.
8. Metabolic labeling can be performed in 12-well plates as well, however, robust cell counts are more easily achieved when the cell line can undergo multiple doublings, which rapidly outpaces the space available in 12-well culture.

Acknowledgments

This work was supported by funds from the National Institutes of Health (5T32GM007347, A.T.H.) and (1R01GM12440401, J.D. Y).

References

1. Berridge MJ, Irvine RF (1989) Inositol phosphates and cell signalling. Nature 341 (6239):197–205. https://doi.org/10.1038/341197a0
2. Hatch AJ, York JD (2010) SnapShot: Inositol phosphates. Cell 143(6):1030–1030.e1031. https://doi.org/10.1016/j.cell.2010.11.045
3. Majerus PW (1992) Inositol phosphate biochemistry. Ann Rev Biochem 61:225–250. https://doi.org/10.1146/annurev.bi.61.070192.001301
4. Bennett M, Onnebo SM, Azevedo C, Saiardi A (2006) Inositol pyrophosphates: metabolism and signaling. Cell Mol Life Sci 63 (5):552–564. https://doi.org/10.1007/s00018-005-5446-z
5. Chakraborty A, Kim S, Snyder SH (2011) Inositol pyrophosphates as mammalian cell signals. Sci Signal 4(188):re1. https://doi.org/10.1126/scisignal.2001958
6. Lee YS, Mulugu S, York JD, O'Shea EK (2007) Regulation of a cyclin-CDK-CDK inhibitor complex by inositol pyrophosphates. Science 316(5821):109–112. https://doi.org/10.1126/science.1139080
7. Pohlmann J, Fleig U (2010) Asp1, a conserved 1/3 inositol polyphosphate kinase, regulates the dimorphic switch in *Schizosaccharomyces pombe*. Mol Cell Biol 30(18):4535–4547. https://doi.org/10.1128/MCB.00472-10
8. Seeds AM, Sandquist JC, Spana EP, York JD (2004) A molecular basis for inositol polyphosphate synthesis in *Drosophila melanogaster*. J Biol Chem 279(45):47222–47232. https://doi.org/10.1074/jbc.M408295200
9. Tsui MM, York JD (2010) Roles of inositol phosphates and inositol pyrophosphates in development, cell signaling and nuclear processes. Adv Enzyme Regul 50(1):324–337. https://doi.org/10.1016/j.advenzreg.2009.12.002
10. York SJ, Armbruster BN, Greenwell P, Petes TD, York JD (2005) Inositol diphosphate signaling regulates telomere length. J Biol Chem 280(6):4264–4269. https://doi.org/10.1074/jbc.M412070200
11. Blind RD, Suzawa M, Ingraham HA (2012) Direct modification and activation of a nuclear receptor-PIP(2) complex by the inositol lipid kinase IPMK. Sci Signal 5(229):ra44. https://doi.org/10.1126/scisignal.2003111
12. Cantley LC (2002) The phosphoinositide 3-kinase pathway. Science 296 (5573):1655–1657. https://doi.org/10.1126/science.296.5573.1655
13. Dyson JM, Fedele CG, Davies EM, Becanovic J, Mitchell CA (2012) Phosphoinositide phosphatases: just as important as the kinases. Subcell Biochem 58:215–279. https://doi.org/10.1007/978-94-007-3012-0_7
14. Hansen SB (2015) Lipid agonism: The PIP2 paradigm of ligand-gated ion channels. Biochimica et Biophysica Acta 1851(5):620–628. https://doi.org/10.1016/j.bbalip.2015.01.011

15. Kim YJ, Hernandez ML, Balla T (2013) Inositol lipid regulation of lipid transfer in specialized membrane domains. Trends Cell Biol 23(6):270–278. https://doi.org/10.1016/j.tcb.2013.01.009
16. Skwarek LC, Boulianne GL (2009) Great expectations for PIP: phosphoinositides as regulators of signaling during development and disease. Develop Cell 16(1):12–20. https://doi.org/10.1016/j.devcel.2008.12.006
17. Vanhaesebroeck B, Stephens L, Hawkins P (2012) PI3K signalling: the path to discovery and understanding. Nat Rev Mol Cell Biol 13 (3):195–203. https://doi.org/10.1038/nrm3290
18. Billcliff PG, Lowe M (2014) Inositol lipid phosphatases in membrane trafficking and human disease. Biochem J 461(2):159–175. https://doi.org/10.1042/BJ20140361
19. Eramo MJ, Mitchell CA (2016) Regulation of PtdIns(3,4,5)P3/Akt signalling by inositol polyphosphate 5-phosphatases. Biochem Soc Trans 44(1):240–252. https://doi.org/10.1042/BST20150214
20. Fruman DA, Chiu H, Hopkins BD, Bagrodia S, Cantley LC, Abraham RT (2017) The PI3K pathway in human disease. Cell 170 (4):605–635. https://doi.org/10.1016/j.cell.2017.07.029
21. Liu Y, Bankaitis VA (2010) Phosphoinositide phosphatases in cell biology and disease. Prog Lipid Res 49(3):201–217. https://doi.org/10.1016/j.plipres.2009.12.001
22. Majerus PW, York JD (2009) Phosphoinositide phosphatases and disease. J Lipid Res 50 (Suppl):S249–S254. https://doi.org/10.1194/jlr.R800072-JLR200
23. Pirruccello M, De Camilli P (2012) Inositol 5-phosphatases: insights from the Lowe syndrome protein OCRL. Trends Biochem Sci 37(4):134–143. https://doi.org/10.1016/j.tibs.2012.01.002
24. Stevenson-Paulik J, Chiou ST, Frederick JP, dela Cruz J, Seeds AM, Otto JC, York JD (2006) Inositol phosphate metabolomics: merging genetic perturbation with modernized radiolabeling methods. Methods 39 (2):112–121. https://doi.org/10.1016/j.ymeth.2006.05.012
25. Fujii M, York JD (2005) A role for rat inositol polyphosphate kinases rIPK2 and rIPK1 in inositol pentakisphosphate and inositol hexakisphosphate production in rat-1 cells. J Biol Chem 280(2):1156–1164. https://doi.org/10.1074/jbc.M412006200
26. Otto JC, Kelly P, Chiou ST, York JD (2007) Alterations in an inositol phosphate code through synergistic activation of a G protein and inositol phosphate kinases. Proc Natl Acad Sci U S A 104(40):15653–15658. https://doi.org/10.1073/pnas.0705729104
27. Otto JC, Mulugu S, Fridy PC, Chiou S-T, Armbruster BN, Ribeiro AA, York JD (2007) Biochemical analysis of inositol phosphate kinases. Methods Enzymol 434:171–185. https://doi.org/10.1016/s0076-6879(07)34010-x
28. Otto JC, York JD (2010) Molecular manipulation and analysis of inositol phosphate and pyrophosphate levels in Mammalian cells. Methods Mol Biol 645:47–60. https://doi.org/10.1007/978-1-60327-175-2_3
29. Seeds AM, Bastidas RJ, York JD (2005) Molecular definition of a novel inositol polyphosphate metabolic pathway initiated by inositol 1,4,5-trisphosphate 3-kinase activity in *Saccharomyces cerevisiae*. J Biol Chem 280 (30):27654–27661. https://doi.org/10.1074/jbc.M505089200
30. Stevenson-Paulik J, Odom AR, York JD (2002) Molecular and biochemical characterization of two plant inositol polyphosphate 6-/3-/5-kinases. J Biol Chem 277 (45):42711–42718. https://doi.org/10.1074/jbc.M209112200
31. Stephens LR, Irvine RF (1990) Stepwise phosphorylation of myo-inositol leading to myo-inositol hexakisphosphate in Dictyostelium. Nature 346(6284):580–583. https://doi.org/10.1038/346580a0
32. Frederick JP, Mattiske D, Wofford JA, Megosh LC, Drake LY, Chiou ST, Hogan BL, York JD (2005) An essential role for an inositol polyphosphate multikinase, Ipk2, in mouse embryogenesis and second messenger production. Proc Natl Acad Sci U S A 102 (24):8454–8459. https://doi.org/10.1073/pnas.0503706102

Chapter 8

Back-Pyrophosphorylation Assay to Detect In Vivo InsP_7-Dependent Protein Pyrophosphorylation in Mammalian Cells

Manasa Chanduri and Rashna Bhandari

Abstract

Protein pyrophosphorylation involves the transfer of a high-energy β-phosphate from inositol pyrophosphates, such as diphosphoinositol pentakisphosphate (InsP_7) to phosphorylated serine residues. Over a decade of research has established several proteins, involved in diverse physiological processes, as substrates of InsP_7-mediated pyrophosphorylation. However, the need for detection of this posttranslational modification on endogenous proteins is paramount. "Back-pyrophosphorylation" is a simple technique to test whether a native protein undergoes InsP_7-mediated pyrophosphorylation inside cells. The basis of this technique relies on the fact that a target protein isolated from cells with lower InsP_7 levels exists in a hypo-pyrophosphorylated form as compared to the same protein isolated from cells with normal InsP_7 levels. Hence, when radiolabeled InsP_7 is added to a target protein immunoprecipitated from both these cell types, the hypopyrophosphorylated protein accepts a higher amount of radiolabeled phosphate when compared to the protein isolated from wild-type cells. This chapter provides detailed methods to identify an InsP_7 target protein and conduct a back-pyrophosphorylation assay on a target protein immunoprecipitated from cells with normal versus reduced InsP_7 levels, to confirm its endogenous pyrophosphorylation status.

Key words Protein pyrophosphorylation, Inositol pyrophosphates, InsP_7, Back-pyrophosphorylation, Immunoprecipitation

1 Introduction

Protein pyrophosphorylation is a post-translational modification that is uniquely conferred on proteins by inositol pyrophosphates. The high-energy β-phosphate of inositol pyrophosphates such as InsP_7 can be transferred to proteins in an enzyme-independent manner, in the presence of Mg^{2+} ions [1]. The target site for pyrophosphorylation is a prephosphorylated Ser residue surrounded by Asp/Glu residues [1]. Since the phosphotransfer from InsP_7 results in the formation of a Ser residue linked to two phosphates, this phenomenon is termed "pyrophosphorylation" (Fig. 1).

Gregory J. Miller (ed.), *Inositol Phosphates: Methods and Protocols*, Methods in Molecular Biology, vol. 2091,
https://doi.org/10.1007/978-1-0716-0167-9_8,

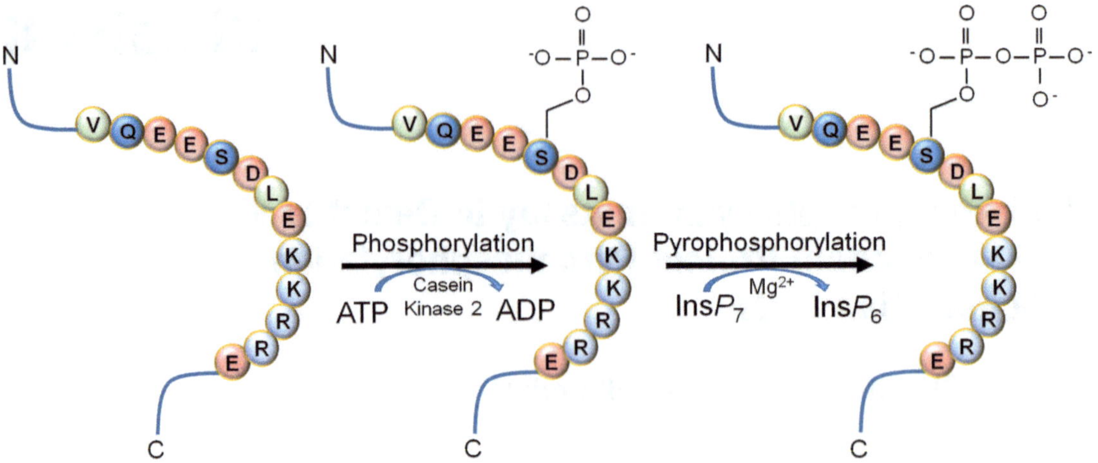

Fig. 1 Protein pyrophosphorylation. A protein containing Ser residues surrounded by acidic amino acid residues (Asp/Glu) is a target site for InsP_7-mediated pyrophosphorylation. The protein is initially phosphorylated by an acidic Ser/Thr kinase such as CK2. InsP_7 then transfers its β-phosphate to the prephosphorylated Ser residue, in the presence of Mg^{2+} ions, to generate a pyrophosphorylated protein. The peptide sequence shown in this figure corresponds to the pyrophosphorylated region on the dynein intermediate chain 2C (IC2C) [5]

A decade of research after the discovery of protein pyrophosphorylation has led to the identification of several protein substrates that are involved in key physiological processes such as ribosome biogenesis, glycolysis, and vesicular trafficking [2–5]. Most substrates of InsP_7-mediated pyrophosphorylation have been identified by overexpression of the target protein followed by in vitro pyrophosphorylation using radiolabeled InsP_7. The detection of this modification on endogenous proteins remains a challenge, as mass spectrometry analysis of cellular proteins has not been able to differentiate between pyrophosphorylation on a single Ser residue and monophosphorylation on two neighboring Ser residues. Although a recent mass spectrometry study combined electron-transfer dissociation with higher-energy collision dissociation to detect pyrophosphorylation on site-specifically modified synthetic pyrophosphopeptides, this method is yet to be extended to endogenous proteins [6].

"Back-pyrophosphorylation" is a simple method to determine whether a target protein is indeed pyrophosphorylated in vivo [4, 5, 7]. The principle of back-pyrophosphorylation relies on the fact that proteins in cells with lower levels of InsP_7 cannot undergo pyrophosphorylation to the same extent as proteins in cells with normal InsP_7 levels. As a result, a target protein isolated from InsP_7 depleted cells, when subjected to pyrophosphorylation using radiolabeled InsP_7, exhibits augmented pyrophosphorylation compared with the same protein isolated from wild-type cells (Fig. 2). This chapter describes methods to identify a target protein for

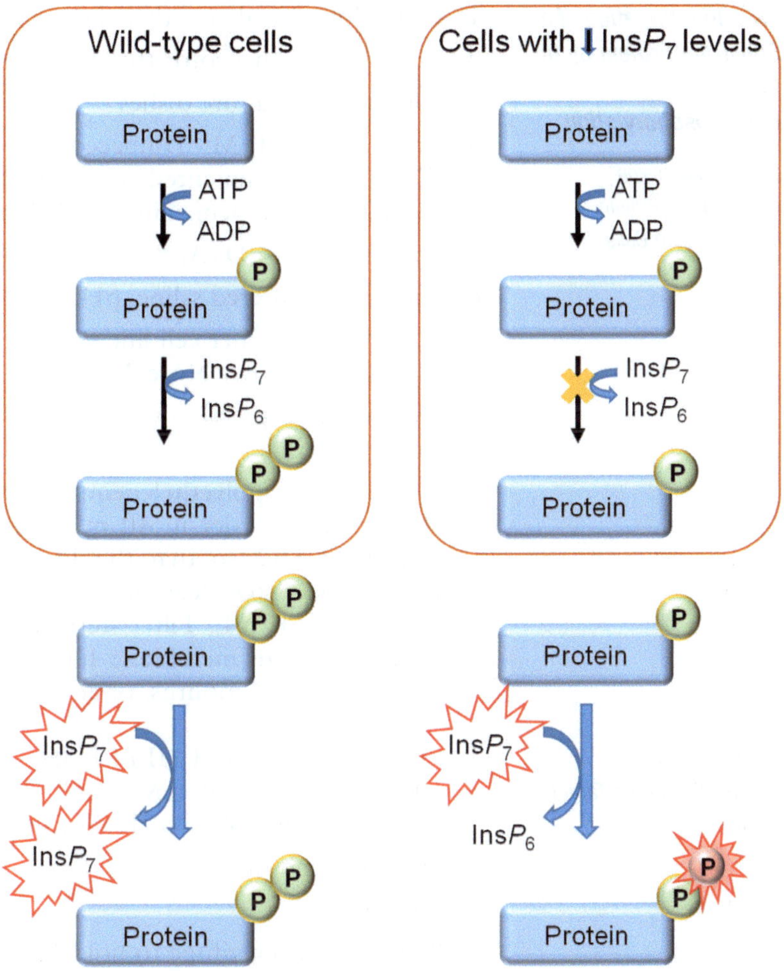

Fig. 2 Back-pyrophosphorylation. An InsP_7 substrate protein exists in a pyrophosphorylated form in wild-type cells containing normal InsP_7 levels, whereas in cells with reduced InsP_7 levels, the protein can be prephosphorylated by a protein kinase, but pyrophosphorylation is reduced or absent. When this protein is isolated from both cell types and treated with radiolabeled InsP_7, the protein isolated from cells with reduced InsP_7 levels exhibits greater pyrophosphorylation when compared to the protein isolated from wild-type cells

pyrophosphorylation by InsP_7, and subsequently determine its endogenous pyrophosphorylation status using the back-pyrophosphorylation assay.

2 Materials

See **Note 1** before preparing solutions described throughout this section.

2.1 Identification of a Substrate Protein by In Vitro Pyrophosphorylation

2.1.1 Expression of the Target Protein in Mammalian Cells

1. Cell culture incubator that can maintain 5% CO_2 and 37 °C.
2. Centrifuge with a rotor that can hold 15 mL conical tubes.
3. 100 mm cell-culture dishes.
4. Complete DMEM: Dulbecco's Modified Eagle's Medium (DMEM) supplemented with 10% FBS, 1 mM L-glutamine, and 100 U/mL penicillin, and 100 μg/mL streptomycin.
5. 0.25% trypsin–EDTA.
6. Phosphate buffered saline, pH 7.4 (PBS).
7. HEK293T or other cell line suitable for transfection and protein expression (*see* **Note 2**).
8. Mammalian expression vector containing cDNA encoding the target protein.
9. 1 mg/mL of polyethyleneimine (PEI) solution: Resuspend 50 mg PEI powder in 40 mL water. Adjust the pH to 7.4 with 1 M HCl so that the PEI powder dissolves and the solution becomes clear, and then add water to make up the volume to 50 mL. Filter-sterilize the PEI solution using a 0.22 μm PVDF membrane filter, and store it in aliquots at 4 °C for up to 2 months, or at −80 °C for long-term storage.

2.1.2 Immunoprecipitation of the Target Protein

1. Lysis Buffer: 25 mM HEPES pH 7.4, 1% NP-40, 0.1% BSA, 100 mM KCl, 150 mM NaCl, 1 mM DTT, protease and phosphatase inhibitor cocktails (Sigma-Aldrich).
2. End-over-end mixer.
3. PBS, pH 7.4.
4. Refrigerated centrifuge to hold 1.5–2 mL tubes.
5. Protein A Sepharose beads (GE Healthcare Life Sciences): Resuspend 50 mg of Protein A Sepharose beads in 15 mL water and incubate overnight at 4 °C. Centrifuge at 5000 × *g* for 5 min to collect the beads. Resuspend the beads in 1 mL water and repeat the centrifugation. Add 1 volume of PBS containing 0.05% sodium azide to the beads, to obtain a 1:1 suspension, which can be stored at 4 °C. Prior to use, wash 30 μL of 1:1 suspension of preswollen Protein A Sepharose beads with 1 mL of PBS pH 7.4, and equilibrate the beads with 1 mL Lysis Buffer. Collect the beads after each step by centrifuging at 5000 × *g* for 5 min at 4 °C. Aspirate the Lysis Buffer and use the beads (~15 μL) for immunoprecipitation.
6. Antibody against the protein of interest or the epitope tag.
7. PBS, pH 7.4, containing protease and phosphatase inhibitor cocktails.

2.2 Back-Pyrophosphorylation Assay to Detect In Vivo Pyrophosphorylation

2.2.1 Immunoprecipitation of Endogenous Protein from $Ip6k1^{+/+}$ and $Ip6k1^{-/-}$ MEFs

1–6. Same as **items 1–6** in Subheading 2.1.1.

7. *Ip6k1*$^{+/+}$ and *Ip6k1*$^{-/-}$ mouse embryonic fibroblasts (MEFs) [8] (*see* **Note 5**).

8–14. Same as **items 1–7** in Subheading 2.1.2.

2.2.2 InsP_7-Mediated Pyrophosphorylation of Immunoprecipitated Proteins

1. Acrylic shielding for protection from beta-particles emitted by ^{32}P.
2. 5[β-^{32}P]InsP_7, prepared as described previously [9] (*see* **Notes 6** and **7**).
3. 10× Pyrophosphorylation Buffer: 250 mM HEPES pH 7.4, 500 mM NaCl, 60 mM $MgCl_2$, 10 mM DTT.
4. Thermal mixer such as ThermoMixer (Eppendorf) to maintain 37 °C while mixing samples.
5. LDS sample buffer 4× (ThermoFisher Scientific).
6. Heating block at 95 °C.
7. A suitable prestained protein molecular weight standard.
8. NuPAGE® Novex 4–12% Bis-Tris gels 1.0 mm × 10-well (ThermoFisher Scientific).
9. 1× MES buffer: Dilute 40 mL of 20× MES buffer (NuPAGE® MES SDS Running Buffer (20×), ThermoFisher Scientific) with water to a final volume of 800 mL.
10. XCell *SureLock*® gel electrophoresis system (ThermoFisher Scientific).
11. Western blot apparatus such as Mini Trans-Blot® Electrophoretic Transfer Cell (Bio-Rad), or an equivalent wet or semidry protein transfer apparatus.
12. PVDF membrane for protein transfer.
13. Western Blotting Buffer: Dissolve 3.04 g of Tris-base and 14 g of glycine in 600 mL water. Add 200 mL methanol and make the volume up to 1 L with water.
14. Geiger counter to monitor radioactivity.
15. Phosphorimager screen and exposure cassette for autoradiography.
16. Phosphorimager scanner such as Typhoon FLA-9500 (GE Healthcare Life Sciences).
17. Suitable primary antibody, and HRP-conjugated secondary antibody for protein detection by immunoblotting.

18. Chemiluminescence detection reagent such as Amersham ECL Prime Western Blotting Detection Reagent (GE Healthcare Life Sciences).
19. X-ray film or a Chemiluminescence imaging system such as FluorChem E (ProteinSimple).
20. Image quantification software, such as ImageJ, for quantifying the extent of pyrophosphorylation on the target protein(s).

3 Methods

When a substrate is identified as a potential target for InsP_7-mediated pyrophosphorylation, the back-pyrophosphorylation assay should be performed to test whether this modification occurs inside cells. This assay has been successfully used to demonstrate that the β subunit of the human adaptor protein complex AP3 (AP3B1), and the intermediate chain of the mouse dynein motor (IC2C), undergo pyrophosphorylation in vivo [4, 5]. The methods below first describe the identification of a mammalian pyrophosphorylation substrate protein, and then confirm its endogenous pyrophosphorylation in mammalian cells using the back-pyrophosphorylation assay.

3.1 Identification of a Substrate Protein by In Vitro Pyrophosphorylation

Potential substrate proteins for pyrophosphorylation by InsP_7 have been identified in three ways: (1) from phenotype to candidate protein, (2) amino acid sequence analysis, and (3) affinity toward a Mg^{2+}-InsP_7 resin. An example of the first strategy is the identification of the dynein intermediate chain as an InsP_7 substrate, arising from the demonstration of perturbed dynein-dependent transport in cells with lower levels of InsP_7 [5]. In the second approach, a protein of interest containing one or more acidic phosphoserine motifs may be tested for InsP_7-mediated pyrophosphorylation, as shown in the case of AP3B1 [4]. To determine whether InsP_7 substrates possess a typical consensus sequence motif, we performed a sequence alignment of proteins on which the site(s) of pyrophosphorylation have been localized (Fig. 3). This alignment reveals a defined consensus composition of multiple Ser/Glu/Asp residues, instead of a linear sequence motif, suggesting that any region with this sequence composition may be tested for pyrophosphorylation. In the third strategy, a subset of yeast proteins that bound to immobilized InsP_7 in the presence of Mg^{2+} ions were found to be pyrophosphorylatable by InsP_7 [6]. Once a potential substrate protein is identified, it may be tested for in vitro pyrophosphorylation by any of the following methods:

(a) Overexpression in yeast: The full-length protein or a region of interest as a fusion to GST may be expressed in *S. cerevisiae*

```
AP3B1I       ----------EDSSDSSSDSESESGSESGEQ
AP3B1II      --------NEDSSEDSSSEQDSESGRESG--
AP3B1III     ---------SSSIEDSSSDSESESEPESES-
NOPP140*     ---------ETSSSDSSEDSSEEEDKAQGL-
RPA34        ---------HHDDEEESSEKKKKKKEKR---
RPA43        SEIVYEENTSESNDGESSDSD----------
RPA190       ---------EGIDSDKESDSDSEDEDVDMN-
IC2C         -------KEAAVSVQEESDLEKKRREAE---
GCR1         -------NSKKVNEDSPSSSSKLAERPR---
NSR1         ----SESESESSSSSSSDSESSSS------
                               •  • •  •
```

Fig. 3 Alignment of known pyrophosphorylation substrate proteins. Sequence alignment of pyrophosphorylated regions of known InsP_7 substrate proteins using CLUSTAL Omega [19, 20]. The sequences correspond to: three regions in human AP3B1 (Uniprot ID: O00203) [4]; mouse NOPP140∗ (GenBank accession no. NP_941035) [1]; *S. cerevisiae* RPA34, RPA43 and RPA190 (Uniprot IDs: P47006, P46669 and P10964) [2]; mouse dynein intermediate chain 2C (IC2C; Uniprot ID: O88487) [5]; *S. cerevisiae* GCR1 (Uniprot ID: P07261) [3]; *S. cerevisiae* NSR1 (Uniprot ID: P27476) [1, 7]. "S" indicates the localized site(s) of pyrophosphorylation on each protein. ∗The sequence of NOPP140 [1] shown here is currently annotated as a predicted pseudogene (Gm5124) located on chromosome X. The corresponding gene, located on chromosome 19, encodes NOLC1 (NOPP140) (Uniprot ID: E9Q5C9), which has a similar amino acid sequence in this region (ETSSSDSSEDSSEDEDKKAQGL), and is likely to be pyrophosphorylated by InsP_7

[10]. The protein may be purified from the yeast extract by affinity purification on glutathione-agarose beads. Alternatively, the yeast extract containing the GST fusion protein may be used directly, along with a control extract from yeast expressing GST alone. The purified protein or yeast extract may be incubated with radiolabeled InsP_7 as described previously [10].

(b) Overexpression in *E. coli*: The full-length protein or a region of interest as a fusion to GST may be expressed in the *E. coli* strain BL21(DE3). The protein may be purified from the bacterial lysate by affinity purification using glutathione-agarose beads, prephosphorylated using the protein kinase CK2 and ATP, followed by pyrophosphorylation using radiolabeled InsP_7 as described previously [10].

(c) Overexpression in a mammalian cell line, as described in Subheading 3.1.1.

3.1.1 Expression of the Target Protein in Mammalian Cells

Subclone the cDNA expressing the target protein or the region of interest into a mammalian expression vector. We recommend using a plasmid carrying a strong promoter, and incorporating a tag fused to the candidate protein to enable its immunoprecipitation. For example, we subcloned the cDNA encoding mouse dynein intermediate chain 2C (IC2C) (NCBI accession number: NM_010064), N-terminally tagged with the V5 epitope, into the plasmid pCDNA3.1(+), and expressed the protein in HEK293T cells.

1. Culture a suitable mammalian cell line, preferably HEK293T (*see* **Note 2**), in a 100 mm dish in 5% CO_2 at 37 °C, until the cells are 40–60% confluent.
2. Transfect the cells with the plasmid encoding the protein or fragment of interest using a transfection reagent such as polyethyleneimine (PEI). For transfection with PEI, mix 8 μg of the plasmid DNA with 24 μL of 1 mg/mL PEI solution in serum- and antibiotic-free DMEM, and incubate at room temperature for 10 min. Wash the cells with PBS and add fresh cell culture medium. Add the transfection reagent–DNA mix to the dish and incubate the cells in 5% CO_2 at 37 °C for 24–48 h.
3. Harvest the cells by trypsinization using 1 mL of 0.25% trypsin–EDTA solution and pellet the cells at 400 × *g* for 3 min at room temperature. Resuspend the cell pellet in PBS and centrifuge at 400 × *g* for 3 min at room temperature to obtain a washed cell pellet.

3.1.2 Immunoprecipitation of the Target Protein

1. To lyse the cells, resuspend the washed cell pellet in 500 μL of cold Lysis Buffer (*see* Subheading 2.1.2) and incubate for 30 min at 4 °C by end-over-end mixing.
2. Centrifuge the cell lysate at 12,000 × *g* for 10 min at 4 °C to remove debris.
3. Collect the supernatant and preclear the lysate by incubating it with 15 μL Protein A Sepharose beads (prepared as described in Subheading 2.1.2) at 4 °C for 1 h with end-over-end mixing.
4. Centrifuge the lysate at 5000 × *g* for 10 min to pellet the Protein A Sepharose beads and collect the supernatant.
5. Add the requisite amount of antibody that recognizes the target protein, or the epitope tagged to the protein, and incubate overnight at 4 °C with end-over-end mixing (*see* **Note 3**).
6. Add the protein–antibody complex to 15 μL Protein A Sepharose beads (prepared as described in Subheading 2.1.2) and incubate at 4 °C for 2 h to immunoprecipitate the target protein.
7. Centrifuge the beads at 5000 × *g* for 10 min and discard the supernatant.
8. Wash the beads twice in PBS containing protease and phosphatase inhibitors and centrifuge at 5000 × *g* for 10 min.
9. Proceed with the pyrophosphorylation assay using radiolabeled InsP_7 (Subheading 3.2.2, **steps 1–9**) (Fig. 4).

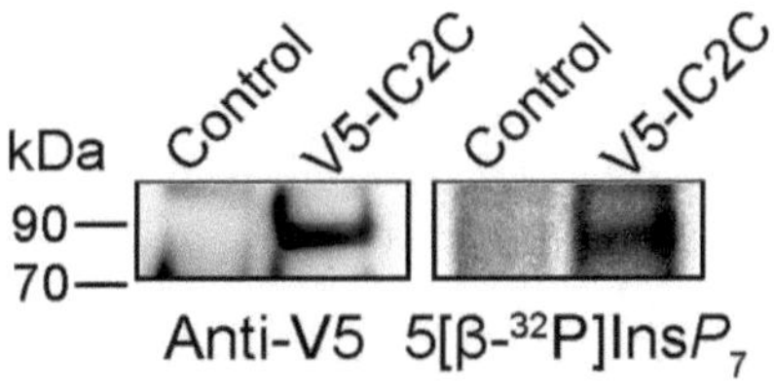

Fig. 4 In vitro pyrophosphorylation of dynein IC2C. V5-tagged dynein intermediate chain 2C (V5-IC2C) was overexpressed in HEK293T cells (*see* Subheading 3.1.1). The protein was immunoprecipitated, as described in Subheading 3.1.2, using an antibody against the V5 epitope (ThermoFisher Scientific, R960-25). HEK293T cells transfected with the empty vector, pCDNA3.1(+), served as a negative control. The immunoprecipitated proteins were incubated with 5 μCi of 5[β-^{32}P]InsP_7 as described in Subheading 3.2.2, resolved on a 4–12% gradient gel, and transferred to a PVDF membrane. InsP_7-mediated pyrophosphorylation was detected by autoradiography (right panel) and the protein was detected by immunoblotting with the anti-V5 antibody (left panel)

3.2 Back-Pyrophosphorylation Assay to Detect In Vivo Pyrophosphorylation

Once an InsP_7 target protein is identified, its endogenous pyrophosphorylation can be confirmed by a back-pyrophosphorylation assay. In this assay, the native protein is isolated from cells containing normal or reduced levels of InsP_7 and subjected to pyrophosphorylation using radiolabeled InsP_7. Native protein isolated from cells lacking InsP_7 exists in a hypo-pyrophosphorylated state when compared to the protein isolated from wild-type cells (*see* **Note 4**). Therefore, when radiolabeled InsP_7 is added to both these proteins, the hypo-pyrophosphorylated protein will accept more phosphate from InsP_7 compared with the protein from wild-type cells (Fig. 2).

This section describes the back-pyrophosphorylation of a native mammalian protein isolated from *Ip6k1*$^{+/+}$ or *Ip6k1*$^{-/-}$ mouse embryonic fibroblasts (MEFs). As *Ip6k1*$^{-/-}$ MEFs show a 70% reduction in the level of InsP_7 compared with *Ip6k1*$^{+/+}$ MEFs, a higher level of phosphotransfer is expected upon back-pyrophosphorylation of the protein from *Ip6k1*$^{-/-}$ MEFs. Instead of MEFs, any other cell line pair with significantly altered InsP_7 levels may be used (*see* **Note 5**). Back-pyrophosphorylation of the protein of interest may also be tested by expressing it in wild-type and *kcs1*Δ *S. cerevisiae*, as previously demonstrated with NSR1 and AP3B1 proteins [4, 7].

3.2.1 Immunoprecipitation of Endogenous Protein from *Ip6k1*$^{+/+}$ and *Ip6k1*$^{-/-}$ MEFs

1. Culture *Ip6k1*$^{+/+}$ and *Ip6k1*$^{-/-}$ MEFs in complete DMEM (*see* Subheading 2.2.1) in 100 mm dishes in 5% CO_2 at 37 °C, until the cells are 80% confluent.
2. Harvest the cells by trypsinization using 1 mL of 0.25% trypsin–EDTA solution and pellet the cells at 400 × *g* for 3 min at room temperature. Resuspend the cell pellet in PBS and centrifuge at

400 × *g* for 3 min at room temperature to obtain a washed cell pellet.

3–10. For cell lysis and immunoprecipitation of the target protein, follow **steps 1–8** described in Subheading 3.1.2.

3.2.2 InsP$_7$-Mediated Pyrophosphorylation of Immunoprecipitated Proteins

1. Set up a pyrophosphorylation reaction using 15 μL of Protein A Sepharose beads containing the immunoprecipitated protein (obtained in Subheading 3.1.2 or 3.2.1), and 3 μL of 10× Pyrophosphorylation Buffer (*see* Subheading 2.2.2). Add 5–7 μCi of 5[β-^{32}P]InsP_7 to the reaction mix (*see* **Notes 6–8**). Add water to bring the reaction volume to 30 μL.
2. Incubate the reaction at 37 °C for 15 min with mixing at 350 rpm. Gently tap the tubes occasionally to ensure that the beads do not settle.
3. Add 10 μL of 4× LDS sample buffer to the reaction mix and heat at 95 °C for 5 min.
4. Load the protein samples on a 4–12% NuPAGE gradient gel and resolve the proteins in 1× MES buffer at 150 V for 90 min, placing the gel apparatus in an ice bath. Excise the lowest portion of the gel to remove the dye-front and proteins less than 3 kDa (*see* **Note 9**).
5. Rinse the gel quickly in water and transfer the proteins to a PVDF membrane by Western blotting. Most proteins in the range 10–180 kDa can be transferred to PVDF by wet transfer for 2 h at 200 mA.
6. Air-dry and expose the membrane to a phosphorimager screen (*see* **Note 10**). Scan the screen at 50 μm resolution on a phosphorimager to detect radioactive protein bands (*see* **Note 11**).
7. Rehydrate the membrane by briefly dipping it in methanol and proceed with immunoblotting using an antibody that recognizes your protein of interest. We recommend using TrueBlot® secondary antibodies (Rockland Immunochemicals Inc.) to avoid interference from heavy or light chains of the antibody used during immunoprecipitation.
8. Detect the protein bands by chemiluminescence and record the image of the immunoblot along with the prestained protein markers using X-ray film or a Chemiluminescence imaging system.
9. Align the autoradiogram and the immunoblot (*see* **Note 11**).
10. Use an image quantification software such as ImageJ to quantify the intensity of bands obtained in the autoradiogram and immunoblot. Normalize the pyrophosphorylation signal from the autoradiogram to the corresponding protein content in the immunoblot. Compare these normalized values for the extent

of pyrophosphorylation in proteins isolated from *Ip6k1*$^{+/+}$ and *Ip6k1*$^{-/-}$ MEFs. If the relative level of pyrophosphorylation is higher in the protein obtained from *Ip6k1*$^{-/-}$ MEFs compared with protein from *Ip6k1*$^{+/+}$ MEFs, it is indicative of endogenous pyrophosphorylation of the protein [4, 5].

4 Notes

1. All buffers should be prepared with ultra-pure deionized water and stored at 4 °C, unless otherwise stated.
2. Ensure that all cell lines are authenticated by short tandem repeat profiling [11].
3. For immunoprecipitation of an overexpressed candidate protein, the corresponding empty vector may serve as a negative control (*see* Fig. 4). Alternatively, when an endogenous candidate protein is tested, normal IgG may be used as a negative control.
4. Before proceeding for back-pyrophosphorylation, it is advisable to check whether the prephosphorylation status of the protein is unchanged in cells with reduced InsP_7 levels. This may be examined either directly by phosphoprotein analysis or indirectly by the back-phosphorylation assay [5].
5. Any cell line pair containing significantly different levels of inositol pyrophosphates may be used to immunoprecipitate the target protein for a back-pyrophosphorylation assay. The best strategy to lower inositol pyrophosphate levels in the cell line of choice is to deplete IP6K1 or IP6K2, as these IP6K isoforms are ubiquitously expressed [12]. The depletion of inositol pyrophosphates should be confirmed by radiolabeling cells with [^{3}H]-*myo*-inositol and monitoring the levels of individual inositol phosphates by HPLC [13–15]. The human colon carcinoma cell line, HCT116, has high levels of inositol pyrophosphates [15, 16], and depletion of IP6K1 or IP6K2 have been shown to significantly lower the levels of InsP_7 in this cell line [14, 17, 18].
6. Take appropriate precautions when handling radioactivity, and always follow the radiation safety guidelines and good laboratory practice rules laid down by your laboratory.
7. For the synthesis of 5[β-^{32}P]InsP_7, we modified the reaction set-up described in the protocol "Synthesis of InsP7 by the Inositol Hexakisphosphate Kinase 1 (IP6K1)," by the Saiardi laboratory [9]. We set up eight reactions of 25 μL each, containing 200 μM IP_6, 40 μM ATP, 350–400 μCi [γ-^{32}P]ATP, and 1 μg His-IP6K1, keeping other reaction conditions the

same. With this method, we obtain 50–100 μCi of 5[β-^{32}P]InsP_7 with a specific activity of approximately 300 Ci/mmole.

8. The final concentration of InsP_7 in the reaction mix should be at least 0.5 μM, to mimic intracellular conditions. Adding 5 μCi of 5[β-^{32}P]InsP_7 (specific activity 300 Ci/mmole) in a 30 μL pyrophosphorylation reaction will achieve a final InsP_7 concentration of approximately 0.5 μM.
9. The lowest portion of the gel, containing unreacted InsP_7, should be excised to ensure that there is no background signal from radiolabeled InsP_7 during detection of the pyrophosphorylated protein. Peptides less than 2 kDa are difficult to test for InsP_7-mediated pyrophosphorylation due to comigration with InsP_7.
10. Prior to setting up exposure of the blot to the phosphorimager screen, monitor the radioactivity (in counts per second) on the membrane using a handheld Geiger counter. The exposure time of the blot to the phosphorimager screen may vary from hours to days, depending on the radioactive counts; the lower the counts, the longer the exposure time.
11. To align the radiolabeled protein bands detected by the phosphorimager with the protein bands on the immunoblot, trace an outline of the blot on a transparent sheet and mark the positions of the prestained protein markers. Overlay the blot outline on a 1:1 print of the phosphorimager scan and mark the positions of the radiolabeled proteins on the transparent sheet. Overlay this trace on a 1:1 print of the immunoblot and align them according to the prestained protein markers, to assess whether the radiolabeled protein bands overlap with the protein bands on the immunoblot.

Acknowledgments

We thank Swarna G. Thota, C.P. Unnikannan, and R. Manorama for generating radiolabeled InsP_7. We thank all members of the Laboratory of Cell Signalling for helpful comments.

References

1. Bhandari R, Saiardi A, Ahmadibeni Y et al (2007) Protein pyrophosphorylation by inositol pyrophosphates is a posttranslational event. Proc Natl Acad Sci U S A 104:15305–15310. https://doi.org/10.1073/pnas.0707338104
2. Thota SG, Unnikannan CP, Thampatty SR et al (2015) Inositol pyrophosphates regulate RNA polymerase I-mediated rRNA transcription in *Saccharomyces cerevisiae*. Biochem J 466:105–114. https://doi.org/10.1042/bj20140798
3. Szijgyarto Z, Garedew A, Azevedo C et al (2011) Influence of inositol pyrophosphates on cellular energy dynamics. Science 334:802–805. https://doi.org/10.1126/science.1211908
4. Azevedo C, Burton A, Ruiz-Mateos E et al (2009) Inositol pyrophosphate mediated

pyrophosphorylation of AP3B1 regulates HIV-1 Gag release. Proc Natl Acad Sci U S A 106:21161–21166. https://doi.org/10.1073/pnas.0909176106

5. Chanduri M, Rai A, Malla AB et al (2016) Inositol hexakisphosphate kinase 1 (IP6K1) activity is required for cytoplasmic dynein-driven transport. Biochem J 473:3031–3047. https://doi.org/10.1042/bcj20160610
6. Penkert M, Yates LM, Schümann M et al (2017) Unambiguous identification of serine and threonine pyrophosphorylation using neutral-loss-triggered electron-transfer/higher-energy collision dissociation. Anal Chem 89:3672–3680. https://doi.org/10.1021/acs.analchem.6b05095
7. Saiardi A, Bhandari R, Resnick AC et al (2004) Phosphorylation of proteins by inositol pyrophosphates. Science 306:2101–2105. https://doi.org/10.1126/science.1103344
8. Bhandari R, Juluri KR, Resnick AC et al (2008) Gene deletion of inositol hexakisphosphate kinase 1 reveals inositol pyrophosphate regulation of insulin secretion, growth, and spermiogenesis. Proc Natl Acad Sci U S A 105:2349–2353. https://doi.org/10.1073/pnas.0712227105
9. Azevedo C, Burton A, Bennett M et al (2010) Synthesis of InsP7 by the inositol hexakisphosphate kinase 1 (IP6K1). In: Barker CJ (ed) Inositol phosphates and lipids: methods and protocols, Methods in molecular biology, vol 645. Humana Press, New York, pp 73–85. https://doi.org/10.1007/978-1-60327-175-2_5
10. Werner JK Jr, Speed T, Bhandari R (2010) Protein pyrophosphorylation by diphosphoinositol pentakisphosphate (InsP7). In: Barker CJ (ed) Inositol phosphates and lipids: methods and protocols, Methods in molecular biology, vol 645. Humana Press, New York, pp 87–102. https://doi.org/10.1007/978-1-60327-175-2_6
11. Almeida JL, Cole KD, Plant AL (2016) Standards for cell line authentication and beyond. PLoS Biol 14:e1002476. https://doi.org/10.1371/journal.pbio.1002476
12. Uhlen M, Fagerberg L, Hallstrom BM et al (2015) Proteomics. Tissue-based map of the human proteome. Science 347:1260419. https://doi.org/10.1126/science.1260419
13. Barker CJ, Illies C, Berggren PO (2010) HPLC separation of inositol polyphosphates. In: Barker CJ (ed) Inositol phosphates and lipids: methods and protocols, Methods in molecular biology, vol 645. Humana Press, New York, pp 21–46. https://doi.org/10.1007/978-1-60327-175-2_2
14. Jadav RS, Kumar D, Buwa N et al (2016) Deletion of inositol hexakisphosphate kinase 1 (IP6K1) reduces cell migration and invasion, conferring protection from aerodigestive tract carcinoma in mice. Cell Signal 28:1124–1136. https://doi.org/10.1016/j.cellsig.2016.04.011
15. Gu C, Wilson MS, Jessen HJ et al (2016) Inositol pyrophosphate profiling of two HCT116 cell lines uncovers variation in InsP8 levels. PLoS One 11:e0165286. https://doi.org/10.1371/journal.pone.0165286
16. Wilson MS, Bulley SJ, Pisani F et al (2015) A novel method for the purification of inositol phosphates from biological samples reveals that no phytate is present in human plasma or urine. Open Biol 5:150014. https://doi.org/10.1098/rsob.150014
17. Koldobskiy MA, Chakraborty A, Werner JK Jr et al (2010) p53-mediated apoptosis requires inositol hexakisphosphate kinase-2. Proc Natl Acad Sci U S A 107:20947–20951. https://doi.org/10.1073/pnas.1015671107
18. Wilson MS, Jessen HJ, Saiardi A (2019) The inositol hexakisphosphate kinases IP6K1 and -2 regulate human cellular phosphate homeostasis, including XPR1-mediated phosphate export. J Biol Chem 294(30):11597–11608.
19. Sievers F, Wilm A, Dineen D et al (2011) Fast, scalable generation of high-quality protein multiple sequence alignments using Clustal Omega. Mol Syst Biol 7:539. https://doi.org/10.1038/msb.2011.75
20. Goujon M, McWilliam H, Li W et al (2010) A new bioinformatics analysis tools framework at EMBL-EBI. Nucleic Acids Res 38:W695–W699. https://doi.org/10.1093/nar/gkq313

Chapter 9

Analyses of Ligand Binding to IP_3 Receptors Using Fluorescence Polarization

Ana M. Rossi and Colin W. Taylor

Abstract

Fluorescence polarization (FP) can be used to measure binding of a small fluorescent ligand to a larger protein because the ligand rotates more rapidly in its free form than when bound. When excited with plane polarized light, the free fluorescent ligand emits depolarized light, which can be quantified. Upon binding, its rotation is reduced and more of the emitted light remains polarized. This allows FP to be used as a nondestructive assay of ligand binding. Here we describe a fast, high-throughput FP assay to quantify the binding of fluorescently labeled inositol 1,4,5-trisphosphate (IP_3) to N-terminal fragments of the IP_3 receptor. The assay is fast (1–6 h), it avoids use of radioactive materials and when measurements are performed at different temperatures, it can resolve Gibbs free energy ($\Delta G°$), enthalpy ($\Delta H°$), and entropy ($\Delta S°$) changes of ligand binding.

Key words IP_3 receptor, Fluorescence polarization, Ligand binding, Affinity, Thermodynamics

1 Introduction

Molecular interactions underpin all biological activity. Quantitative analysis of ligand–protein interactions is essential to understanding cellular behavior. Although several methods have been applied for quantitative analysis of ligand–protein interactions [1], most studies have used radioligand binding assays. These methods are laborious and time-consuming, they are not amenable to high-throughput analyses, and they use expensive and environmentally unfriendly radioactive materials. A major drawback of radioligand binding assays is the need to separate bound from free ligand, which hampers quantification of low-affinity interactions. Fluorescence polarization (FP) assays provide an alternative approach that avoids these problems. FP measures the change in rotational speed of a fluorescent ligand during its excited lifetime as a consequence of it binding to a larger protein or receptor. A fluorophore is optimally excited (typically in 10^{-15}s) by plane-polarized light in a preferred orientation and then (usually) [2, 3] emits light in the same

Gregory J. Miller (ed.), *Inositol Phosphates: Methods and Protocols*, Methods in Molecular Biology, vol. 2091, https://doi.org/10.1007/978-1-0716-0167-9_9, © Springer Science+Business Media, LLC, part of Springer Nature 2020

orientation. If there is no movement of the molecule during the fluorescent lifetime of the fluorophore (typically ns), then 60% of all emitted light would be detected in the same plane as the exciting light (parallel) and 20% in each of the other two planes (perpendicular) (Fig. 1a). But if the molecule rotates (<ns timescales for small molecules), it will no longer be aligned and when the fluorophore emits its light <60% will be aligned with the source (Fig. 1a). When plane-polarized light is used for excitation, then the intensities of the light detected in the perpendicular ($I_{\perp}$) and parallel planes ($I_{||}$) with respect to the excitation source are related to the polarization (P) and anisotropy (A):

$$P = \frac{I_{||} - I_{\perp}}{I_{||} + I_{\perp}} \tag{1}$$

$$A = \frac{I_{||} - I_{\perp}}{I_{||} + 2I_{\perp}} \tag{2}$$

The denominator in Eq. 2 is the total light ($2I_{\perp}$ to take account of both perpendicular planes); A therefore provides the most algebraically convenient description of FP because anisotropy components can be summed (polarization values cannot) [4]. A fluorescent molecule with the properties described above would then (with no rotation) have:

$$P = \frac{60 - 20}{60 + 20} = 0.5$$

$$A = \frac{60 - 20}{100} = 0.4$$

If the molecule tumbles so that it becomes randomly reoriented during the fluorescent lifetime, then both A and P become 0 (because $I_{||} = I_{\perp}$). With other conditions (viscosity, etc.) constant, size determines whether the molecule can reorient during the fluorescent lifetime [2, 3]. For large molecules, the speed of tumbling is slower than typical fluorescent lifetimes; now a molecule is likely to retain the same orientation throughout the interval between absorbing and emitting a photon. Binding of a fluorescent ligand ($D*$) to a larger protein (R) increases the effective size of the fluorescent complex and it therefore tumbles more slowly, causing more emitted light to remain polarized with respect to the excitation light (Fig. 1b). This allows FP to be used as a nonperturbing assay of ligand binding (Fig. 1a–c).

Inositol 1,4,5-trisphosphate receptors (IP_3R) are intracellular channels, which are expressed in the endoplasmic reticulum of almost, if not all cells [5]. They mediate the initial release of Ca^{2+} from intracellular stores evoked by cell-surface receptors that induce IP_3 formation and contribute to regulating a myriad of

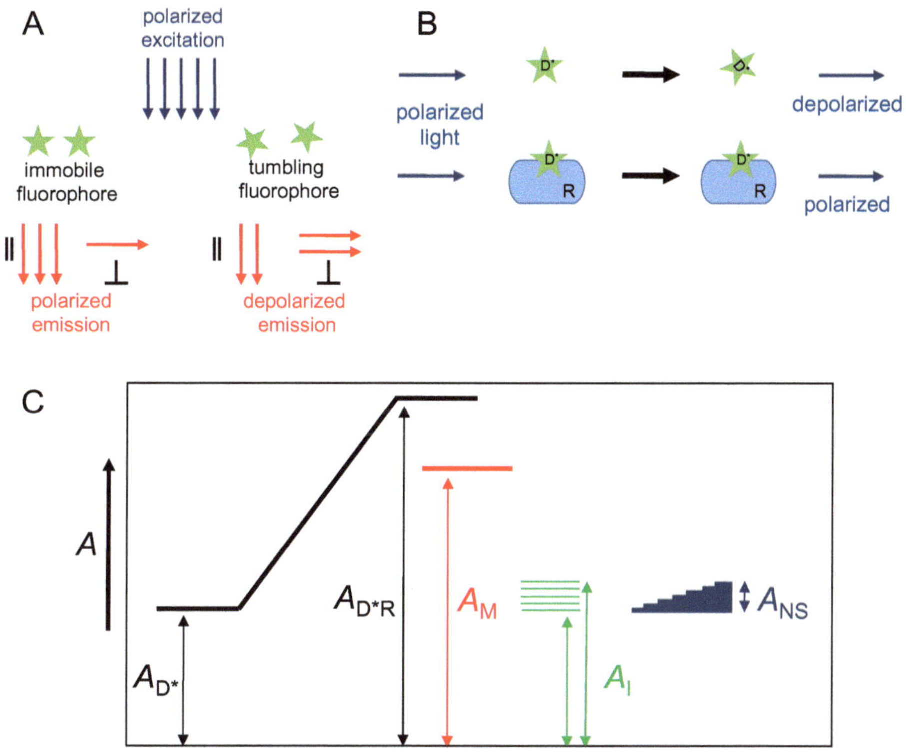

Fig. 1 Use of fluorescence polarization to measure ligand binding. (**a**) When an immobile fluorophore is excited with plane-polarized light, 60% of the emitted light remains polarized with respect to the excitation source (i.e., parallel). If the molecule rotates during its fluorescence lifetime, then less than 60% of the emitted light will remain polarized in the parallel plane and more light will be collected in each of the perpendicular planes [3]. FP measures the emission by a fluorophore in two planes (horizontal and vertical) after excitation with polarized light in one of these planes. As described in Eq. 2, anisotropy (*A*) is defined by the relationships between the intensities of the light detected in the parallel ($I_{\parallel}$) and perpendicular ($I_{\perp}$) planes with respect to the excitation light. For an immobile fluorescent molecule: $A = (60 - 20)/(60 + 40) = 0.4$ [3]. If the molecule rotates and becomes randomly reoriented, $I_{\parallel} = I_{\perp}$ and $A = 0$. For a mixture of fluorescent molecules: $A = \sum_{i=1}^{n} f_i A_i$, where A_i is the anisotropy of each molecule and f_i denotes its fractional fluorescence intensity. As the fraction of reoriented molecules increases, *A* decreases linearly from 0.4 (its maximum value) to 0. (**b**) A small fluorescent ligand (D^*) free in solution rotates rapidly due to its small molecular volume. If D^* is excited with plane-polarized light, it will emit depolarized light and thus will have a low *A* (top). If D^* binds to a large protein (R), its molecular volume will increase and it will therefore rotate less (i.e., the speed of rotation is inversely related to molecular volume [3]). Thus, more of the emitted light will remain polarized with respect to the excitation light, causing *A* to increase (bottom). (**c**) The maximal dynamic range of an FP assay is given by the anisotropy (*A*) of the free D^* (A_{D*}) and bound D^* (A_{DR*}). A_M, the measured *A* for each experimental condition, includes *A* from free D^*, and D^* specifically bound to *R* and nonspecifically bound to other components. A_I (for each experimental condition), is the *A* measured in the presence of a saturating concentration of a competing ligand (*I*) which causes displacement of all D^* specifically bound. A_{NS} (defined as *A* due to nonspecific binding) is calculated from A_I using Eqs. 2–4 and A_S (*A* due to specific binding) using Eq. 5. (Reproduced, with modifications from ref. 1)

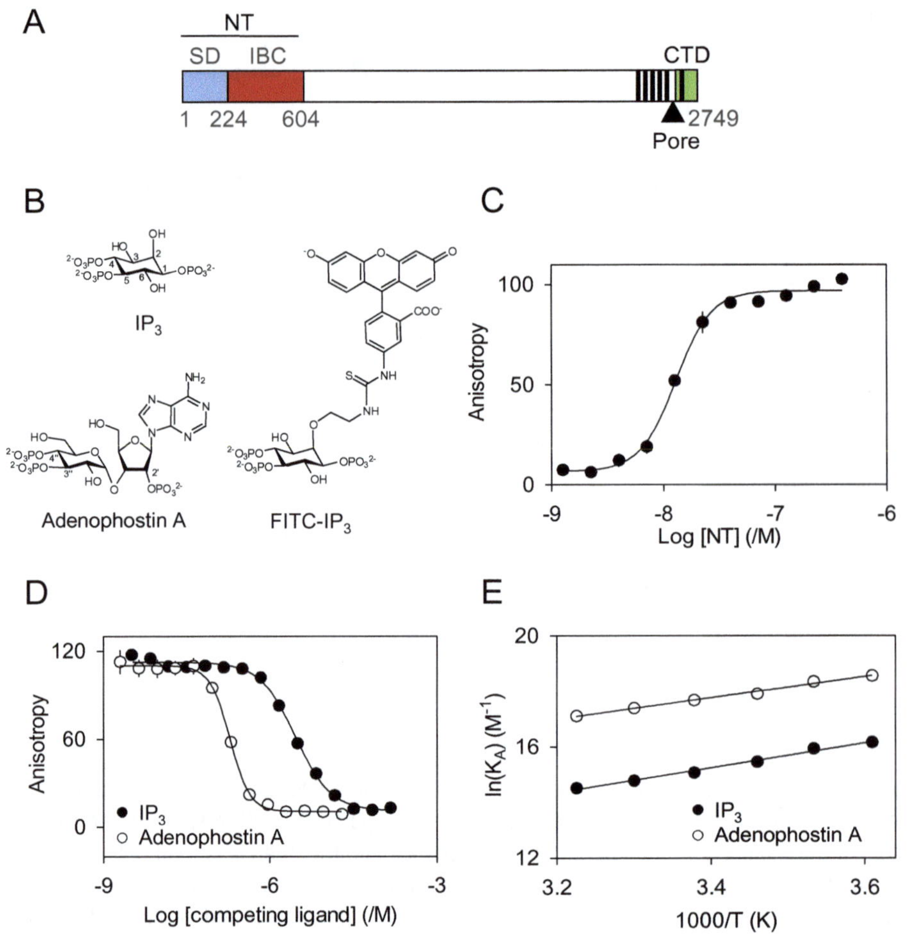

Fig. 2 FP measurements of equilibrium ligand binding to N-terminal fragments of IP_3R (**a**) Cartoon of a single subunit of IP_3R1 showing its key domains: the NT (residues 1-604) which comprises the IBC and SD, 6 transmembrane domains (black lines) and the C-terminal domain (CTD). (**b**) Chemical structures of IP_3, FITC-IP_3 and adenophostin A. (**c**) Anisotropy measurements from a typical FP saturation binding assay using FITC-IP_3 (0.5 nM) and the indicated concentrations of purified NT performed in Ca^{2+}-free CLM at 4 °. (**d**) Typical results from a FP competition binding assay using FITC-IP_3 (0.5 nM), NT (80 nM) and the indicated concentrations of competing ligands in Ca^{2+}-free CLM at 4°. (**e**) Typical van't Hoff plots for IP_3 and adenophostin A binding to the NT. Results (**c–e**) are means ± S.E.M., $n = 3$. (Original data reproduced from ref. 14 with permission from The American Society for Pharmacology and Experimental Therapeutics)

cellular processes [5]. The functional channel is a tetramer of the same or different subunits encoded by three different IP_3R genes (IP_3R1-3) and its opening requires binding of both IP_3 and Ca^{2+} [6]. The IP_3-binding core (IBC, residues 224-604 of IP_3R; Fig. 2a) of each subunit is entirely responsible for binding IP_3 [7]. Residues immediately preceding the IBC (residues 1-223, suppressor domain; SD) decrease the IP_3-binding affinity. Thus, the IBC displays substantially higher affinity for IP_3 than the full-length IP_3R or an N-terminal fragment comprising the SD and IBC (residues

1-604; N-terminal; NT) [8] (Fig. 2a). IP_3 binding initiates channel activation by closing the clam-like IBC and displacing the SD [9, 10], which is essential for channel gating [8, 11]. These conformational changes are probably transmitted directly to the channel pore *via* direct contacts between the IBC and the C-terminal domain (CTD) (Fig. 2a), which is contiguous with the pore [12]. The IBC and NT fragments can be expressed as soluble proteins in bacteria and recognize IP_3 and related ligands with the same specificity as native IP_3R [11, 13]. Here we use IBC and NT fragments of IP_3R1 to demonstrate the utility of FP for analyses of ligand binding to IP_3R [14]. For these analyses we use IP_3 labeled at its 2-position with fluorescein isothiocyanate (FITC-IP_3) because 2-modified analogues of IP_3 bind to the IBC and activate the IP_3R [11]. We initially confirmed the validity of the FP assay by measuring the affinities of FITC-IP_3, IP_3, and adenophostin A (Fig. 2b) to the NT and IBC and the results were comparable to those obtained using conventional radioligand binding assays [14]. Adenophostin A is a metabolically stable, high-affinity agonist of IP_3Rs and residues within the IBC are sufficient to confer the high-affinity binding [13]. Typical results obtained using the FP assay are shown in Fig. 2. These include an equilibrium saturation curve with FITC-IP_3 and increasing concentrations of the NT (Fig. 2c), and equilibrium competition curves using IP_3 or adenophostin A as competing ligands and a fixed concentration of the NT (Fig. 2d). In a typical equilibrium competition binding assay, a single 96-well plate provides sufficient data to determine the K_D for 3 different ligands (using 15 concentrations of each ligand in duplicates). The nondestructive nature of the assay allows analysis of the effects of temperature on ligand binding. The same 96-well plate can be used repeatedly to measure the K_D of the 3 ligands at 6 different temperatures in less than 3 h. These analyses can then provide estimates of the enthalpy (ΔH°) and entropy (ΔS°) changes associated with ligand binding (Fig. 2e). Reusing the same plate for repeated measurements reduces the variability due to plate-to-plate differences and it makes efficient use of precious materials. Similar experiments using radioligand binding assays would have been impracticable due to the need of expensive radioligands, time (i.e., would have taken several days to complete) and the fast off-rates at high temperatures.

Our FP assay, once optimized, provides a rapid (1–3 h) and relatively inexpensive way of measuring ligand binding to IP_3R and its fragments in real time. Since it does not require separation of ligand and protein, it can detect even low-affinity interactions. The assay avoids use of radioactive materials, it is nondestructive and it can resolve Gibbs free energy (ΔG°), enthalpy (ΔH°) and entropy (ΔS°) changes of ligand binding. The protocol is divided into five steps comprising optimization (Subheading 3.1), preparation of plates for equilibrium saturation (Subheading 3.2) or competition (Subheading 3.3) binding assays, anisotropy measurements

(Subheading 3.4) and data analysis (Subheading 3.5). We describe the methods to use with the NT and FITC-IP_3. Other fragments and fluorescent ligands can be used, but the protocol would need to be modified accordingly.

2 Materials

1. Ca^{2+}-free cytosol-like medium (Ca^{2+}-free CLM): 140mM KCl, 20mM NaCl, 2mM $MgCl_2$, 1mM EGTA, and 20mM PIPES, pH 7.0. Use ultrapure water (purified using Milli-Q or similar) and analytical grade reagents. The solution can be stored at 4 °C for several months.
2. "FITC-IP_3" (D-2-*O*-(2-(3-(5-fluoresceinyl)thioureido)ethyl)-*myo*-inositol 1,4,5-trisphosphate). FITC-IP_3 was synthesized and purified as described [14] (*see* **Note 1**).
3. Purified N-terminal fragments of IP_3R. N-terminal fragments of IP_3R1 were expressed in *E. coli* and then purified and quantified (*see* **Note 2**) as previously described [1]. We used the N-terminal (NT, residues 1-604 of IP_3R1) and the IP_3-binding core (IBC, residues 224-604 of IP_3R1), but other fragments or mutant versions could be used.
4. Half-area, black, flat-bottomed, polystyrene 96-well microplates.
5. Inositol 1,4,5-trisphosphate hexapotassium salt (IP_3) and other competing ligands (*see* **Note 3**).
6. Bench-top centrifuge for 96-well plates.
7. Multichannel (8-channel) for using with 96-well plates and tips.
8. Plate reader suited for FP measurements (e.g., PHERAstar, BMG Labtech).
9. Curve-fitting software.

3 Methods

3.1 Optimization of an FP Assay for N-Terminal Fragments of IP_3R (See Note 4)

3.1.1 Assess Whether the Fluorescence Intensity of the Ligand (Here FITC-IP_3) Changes upon Binding to NT (See **Note 5**)

1. Prepare 15 dilutions of NT (with final concentrations ranging between 1 and 300 nM) by serially diluting the purified NT sample ("serially diluted NT").
2. Mix each concentration of purified protein with FITC-IP_3 (final concentration, 0.5 nM) in Ca^{2+}-free CLM.
3. Load duplicate 50-μl samples of each dilution into individual wells of a 96-well plate.
4. Centrifuge the plate (300 × *g*, 2 min, 20 °C).
5. Incubate for 20 min to allow equilibrium to be attained.

6. Measure the intensity of the fluorescence emitted at 520 nm with excitation at 485 nm using a suitable plate reader (e.g., PHERAstar). Note that no measurements of polarized light are required in this step.

3.1.2 Establish the Optimal Concentration of FITC-IP_3 to Be Used in the Assay

1. Prepare serial dilutions of FITC-IP_3 (final concentrations, ~0.1 nM to ~5 nM) in Ca^{2+}-free CLM.
2. Load duplicate 50-μl samples of each dilution into a 96-well plate.
3. Centrifuge the plate (300 × *g*, 2 min, 20 °C).
4. Measure fluorescence intensity in the parallel ($I_{||}$) and perpendicular planes ($I_{\perp}$) and thereby anisotropy (A) (*see* Eq. 2) using the PHERAstar plate reader as described in "Anisotropy measurements."
5. Choose the lowest concentration of FITC-IP_3 (to keep it below its equilibrium dissociation constant, K_D) that allows A to be reliably measured. Here we used 0.5 nM FITC-IP_3.

3.1.3 Assess Whether the Protein Interferes with the Detection of Polarized Light (See ***Note 6****)*

1. Mix "serially diluted NT" (final concentrations, 1–300 nM) with FITC-IP_3 (final concentration, 0.5 nM) and saturating IP_3 (final concentration, 10 μM) in Ca^{2+}-free CLM.
2. Load duplicate 50-μl samples into a 96-well plate.
3. Mix by shaking the plate gently (30 rpm, 20 °C) using the orbital shaking mode in the PHERAstar plate reader (or an orbital shaker).
4. Centrifuge (300 × *g*, 2 min, 20 °C).
5. Incubate for 20 min and then measure $I_{||}$ and $I_{\perp}$ (and thereby A) using the PHERAstar plate reader as described in "Anisotropy measurements."

3.2 Preparation of Plates to Perform FP Equilibrium Saturation Binding Assays

1. Prepare 15 dilutions of the purified NT (with final concentrations ranging between 1 and 300 nM) by serially diluting the purified NT sample ("serially diluted NT").
2. Load a 96-well plate with duplicate 50-μl samples in Ca^{2+}-free CLM of:
   (a) Each dilution of "serially diluted NT" and FITC-IP_3 (final concentration, 0.5 nM) to determine A_M (the measured A, Fig. 1c) for each NT concentration, and A_{D*R} (the anisotropy of bound fluorescent ligand, D^*, Fig. 1c) from the well containing the highest concentration of NT;
   (b) Each dilution of "serially diluted NT" and FITC-IP_3 (final concentration, 0.5 nM), with a saturating concentration of IP_3 (final concentration, 10 μM) to determine A_I (A in the presence of saturating competing ligand, Fig. 1c) and

thus A_{NS} (A due to nonspecific binding of D^*, Fig. 1c) at each NT concentration;

(c) Each dilution of "serially diluted NT" alone to measure background fluorescence ($I_{||}$ and $I_{\perp}$) at each NT concentration. If background fluorescence values are significant, they must be subtracted from the equivalent measurements in the presence of FITC-IP_3 before calculating A;

(d) FITC-IP_3 (final concentration, 0.5 nM) to determine A_{D*} (the anisotropy of free D, Fig. 1c).

Each of these conditions (**a**–**d**) should be included in every plate.

3. Mix by shaking plate (30 rpm, 15 min, 20 °C) using the orbital shaking mode in the PHERAstar plate reader (or an orbital shaker).
4. Wrap plates and samples in foil to minimize exposure to light.
5. Centrifuge plate (300 × *g*, 2 min, 20 °C).
6. Incubate the plate in the dark at the desired temperature (4–37 °C) for 20 min to allow equilibrium to be attained (*see* **Note 7**).
7. Using the PHERAstar plate reader, measure $I_{||}$ and $I_{\perp}$ (and thus A).

3.3 Preparation of Plates to Perform FP Equilibrium Competition Binding Assays

1. Prepare 8 dilutions of competing ligand (e.g., IP_3) with final concentrations ranging between 3 nM and 150 μM) by serially diluting the stock ("serially diluted competing ligand") (*see* **Note 8**).
2. Load a 96-well plate with duplicate 50-μl samples in Ca^{2+}-free CLM containing:

(a) Each dilution of "serially diluted competing ligand," FITC-IP_3 (final concentration, 0.5 nM) and NT (final concentration, 80 nM) to determine A_M at each concentration of competing ligand, and A_I at a saturating concentration of the competing ligand;

(b) FITC-IP_3 alone (final concentration, 0.5 nM) to measure A_{D*};

(c) FITC-IP_3 (final concentration, 0.5 nM) and a saturating concentration of NT (final concentration, 300 nM), to determine A_{D*R} (under these conditions all FITC-IP_3 should be bound) (*see* **Note 9**);

(d) NT (final concentration, 80 nM) to measure background $I_{||}$ and $I_{\perp}$.

3. Mix by shaking plate (30 rpm, 15 min, 20 °C) using the orbital shaking mode in the PHERAstar plate reader (or an orbital shaker).

4. Wrap plates and samples in foil to minimize exposure to light.
5. Centrifuge the plate (300 × *g*, 2 min, 20 °C).
6. Incubate the plate in the dark at the desired temperature (4–37 °C) for 20 min to allow equilibrium to be attained (*see* **Note 7**).
7. Using the PHERAstar plate reader, measure $I_{||}$ and $I_{\perp}$ (and thus A).

3.4 Anisotropy Measurements

Here we describe the methods to use a PHERAstar fluorescence polarization plate reader. Similar equipment from other providers is also suitable, but modified procedures will be required.

1. Open the PHERAstar software and select the FP option. Insert the FP 485/520/520 optic module. Set the following options: 3 cycles of measurement, 300 flashes per cycle, 1 s positioning delay, and reading direction 11.
2. Use the well containing only 50 μl of 0.5 nM FITC-IP_3 to perform autoadjustment of focal height and gain. The "mP Target" value should be set to 25. The focal height for 50 μl of solution in a half-area well should be ~6.3 (*see* **Note 10**).
3. With the above settings take measurements from every well in the plate. The samples are excited with light polarized in the horizontal plane at 485 nm, and emitted light at 520 nm is collected simultaneously in both the horizontal (parallel, $I_{||}$) and vertical (perpendicular, $I_{\perp}$) planes.
4. For repeated measurements of the same plate at different temperatures, both plate reader and plate should be re-equilibrated to the desired temperature. After equilibration and before performing measurements, the plate should be incubated for 20 min at each temperature to allow reactions to attain equilibrium (*see* **Note 11**).

3.5 Data Analysis

Before proceeding with analysis, prepare suitable spreadsheets (e.g., Microsoft Excel) to analyze the data directly imported from the plate reader.

1. For each measurement, subtract the background fluorescence $I_{||}$ and $I_{\perp}$ caused by the presence of protein (here NT) alone. Use these corrected values to calculate the measured A (A_M, *see* Fig. 1c) using Eq. 2.
2. Using Eq. 2, similarly calculate A_I (A determined in the presence of a saturating concentration of a competing ligand) from values of $I_{||}$ and $I_{\perp}$ determined for each NT concentration in the presence of a saturating concentration of competing ligand (IP_3). From the calculated A_I and A_{D*} (A for free D^*, here

FITC-IP$_3$), compute A_{NS} (A due to nonspecific binding, *see* Fig. 1c) at each NT concentration using:

$$A_{NS} = (A_I - A_{D^*})(1 - F_B) \tag{3}$$

where F_B (the fraction of bound D^*) is calculated using:

$$F_B = \frac{A_M - A_{D^*}}{A_{D^*R} - A_{D^*}} \tag{4}$$

and A_{D*R} denotes A for D^* bound to R.

3. For each determination of A_M and its corresponding A_{NS}, calculate A_S (A due to specific binding of FITC-IP$_3$) using:

$$A_S = A_M - A_{NS} \tag{5}$$

4. Use the A_S value (calculated from Eq. 5) directly to plot the saturation binding curve or to calculate the fraction of D^* (fluorescent ligand, here FITC-IP$_3$) bound to R (receptor, here NT) using:

$$\frac{[D^*R]}{[D^*]_T} = \frac{A_S - A_{D*}}{A_{D*\mathrm{R}} - A_{D*}} \tag{6}$$

where $[D^*]_T$ denotes the total concentration of D^*.

5. If $[R]_T$ (the total R concentration) is high enough to ensure minimal depletion of the free R concentration $[R]$ after binding D^* (and thereby $[R]_T \sim [R]$), plot directly A_S against $[R]_T$ to obtain a binding curve. If $[R]_T$ is too low to avoid a significant decrease in $[R]$ as it binds to D^*, $[R]$ should be calculated for each condition before plotting the binding curve using:

$$\frac{[D^*]}{[D^*]_T} = \frac{K_D}{K_D + [R]_T} \tag{7}$$

where K_D is the equilibrium dissociation constant and $[D^*]$ and $[D^*]_T$ are the free and total concentrations of D^*, respectively. In either case, use any suitable curve-fitting program to fit the results to a Hill equation:

$$\alpha = \frac{\left(\frac{[R]}{K_D}\right)^h}{1 + \left(\frac{[R]}{K_D}\right)^h} \tag{8}$$

where α denotes the fraction of D^* specifically bound, and h is the Hill coefficient. From this, determine the $K_D{}^{D*}$, that is K_D of the interaction between D^* and R.

6. Once the $K_D{}^{D*}$ is established, the K_D of any competing ligand ($K_D{}^I$) can be computed (*see* **Note 12**). For equilibrium competition binding measurements, calculate A_M and then A_S from each measurement and from these A_S values compute the total

concentration (I_{50}) of the inhibitor (I) that displaces 50% of D^* specifically bound. From the I_{50} value and the known $[R]_T$, calculate the free concentration of I (IC_{50}) that displaces 50% of specifically bound D^*. At the I_{50}, 50% of all R are occupied by I, hence:

$$IC_{50} = I_{50} - [R]_T/2 \tag{9}$$

7. From A_S value measured at the I_{50} ($A_S{}^{50}$), calculate the $[D^*R]$ at the I_{50} ($[D^*R]_{50}$) using:

$$[D^*R]_{50} = [D^*]_T \frac{A_S^{50} - A_{D*}}{A_{D*R} - A_{D*}} \tag{10}$$

8. The K_D of I for the binding site ($K_D{}^I$) is then calculated using:

$$K_D{}^I = \frac{[D^*R]_{50}\, IC_{50}\; K_D{}^{D*}}{([D^*]_T[R]_T) + [D^*R]_{50}([R]_T - [D^*]_T + [D^*R]_{50} - K_D{}^{D*})} \tag{11}$$

9. From experiments performed at different temperatures, the K_D can be calculated from saturation or competition experiments at each chosen temperature. The following equations define the relationships between K_D, ΔG° (standard Gibbs free energy change), ΔS° (standard entropy change) and ΔH° (standard enthalpy change):

$$\Delta G^\circ = RT \ln K_D \tag{12}$$

$$\Delta G^\circ = \Delta H^\circ - \Delta S^\circ \tag{13}$$

where R is the universal gas constant and T the absolute temperature. If ΔH° is temperature-independent (i.e., the change in heat capacity, $\Delta C = 0$), a van't Hoff plot ($\ln K_D$ versus $1/T$) can be used to obtain $\Delta H^\circ/R$ from the slope. In cases where $\Delta C \neq 0$, the following equation should be used [15]:

$$\Delta G^\circ_T = \Delta H^\circ_{T_o} + \Delta C^\circ_{T_o}(T - T_o) - T\left[\Delta S^\circ_{T_o} + \Delta C^\circ_{T_o} \ln\left(\frac{T}{T_o}\right)\right] \tag{14}$$

where T is the actual absolute temperature, and T_o is the reference absolute temperature (here 296 K).

10. Determine ΔH°, ΔS° and ΔC° by least-squares curve fitting of ΔG° versus T.

4 Notes

1. This protocol is described for FITC-IP_3, but other fluorescent ligands can be used. If using other fluorophores, ensure the excitation and emission wavelengths used are optimal for the chosen fluorophore and that binding to the protein does not affect the fluorescent properties of the ligand (*see* also **Note 5**).
2. Accurate determination of the concentration of functional ligand-binding sites in the purified protein is crucial for all FP experiments. We used conventional radioligand binding assays with ^{3}H-IP_3 to determine the density of binding sites in the purified protein sample [1]. In all analyses, the concentration of purified protein refers to the concentration of functional binding sites.
3. The specific needs of each experiment will determine the need for other ligands (adenophostin A, synthetic analogues of IP_3R, etc. [11]).
4. This protocol is described for the NT. For the IBC, use ~10-fold lower concentrations of functional binding sites (because its affinity for IP_3 is ~10-fold higher than the NT). If using other fragments or mutant fragments, determine their K_D for IP_3 by other methods and adjust the concentration of functional binding sites accordingly.
5. For a fluorescent ligand to be suitable for FP analyses, its rate of depolarization, but not its fluorescence intensity, should change when binding to the protein. If the fluorescence intensity of the ligand changes upon binding the protein, try another fluorescent ligand. This might be achieved by either conjugating the fluorophore to a different position or using a different fluorophore.
6. For all FP measurements, it is crucial to determine whether the protein itself interferes with the detection of polarized light. Here measurements are performed in the presence of a large excess of unlabeled ligand in order to prevent binding of FITC-IP_3 to NT. If attenuation of the emitted light is the same in both planes ($I_{||}$ and $I_{\perp}$), it will not affect measurements of A. Otherwise a correction will be required.
7. The time required for reactions to attain equilibrium will depend on the protein and ligand, their concentrations and the temperature. The timing must be optimized for each set of conditions. If equilibrium has not been attained before performing measurements for a given set of conditions, optimize the timing by measuring the same plate at regular intervals (e.g., every 5 min) to follow the increase in A overtime.

Determine the time at which A reaches a plateau, indicating that equilibrium has been attained.

8. The concentrations of the competing ligand used should cover a range of at least 100-fold either side of its K_D for the protein used.
9. The concentration of purified protein used should be optimized to both avoid wasting valuable material and to maximize the dynamic range. Choose the lowest possible protein concentration to ensure that most fluorescent ligand (here FITC-IP_3) is bound to the protein before addition of competing ligand. We found that protein concentrations that gave A greater than 50% of A_{D*R} (Fig. 1c) worked well.
10. Optimally adjust the gain and focal height to achieve maximal sensitivity.
11. High temperatures might affect the stability of the ligand or the protein (e.g., by causing protein damage or ligand degradation). Verify that the effects of increasing temperature are reversible by measuring the plate at the lowest chosen temperature, then equilibrating the plate at the highest temperature, and repeating the measurement after returning the plate to the lowest temperature. The PHERAstar plate reader is equipped with a temperature-control system. If using a plate reader without temperature control, the plate reader should be housed in a temperature-controlled cabinet.
12. In conventional radioligand binding experiments, the Cheng-Prusoff equation can be used to calculate the K_D from the concentration of drug that displaces 50% of specifically bound radioligand (IC_{50}) [16]. In FP experiments, the relationships between total and free concentrations are more complicated, so the ${K_D}^I$ cannot be calculated using the Cheng-Prusoff equation [17].

Acknowledgments

This work was supported by the Wellcome Trust (101844) and the Biotechnology and Biological Sciences Research Council, UK (BB/P005330/1). Ana M Rossi is a fellow of Queens' College, Cambridge. We thank Zhao Ding, Barry Potter (University of Oxford), and Andrew Riley (University of Bath) for contributing to the development of the FP assay [14].

References

1. Rossi AM, Taylor CW (2011) Analysis of protein-ligand interactions by fluorescence polarization. Nat Protoc 6:365–387
2. Weber G (1952) Polarization of the fluorescence of macromolecules. I. Theory and experimental method. Biochem J 51:145–155
3. Serdyuk I, Zaccia N, Zaccai J (2007) Methods in molecular biophysics. In: Structure, dynamics and function. Cambridge University Press, Cambridge, p 1120
4. Hall MD et al (2016) Fluorescence polarization assays in high-throughput screening and drug discovery: a review. Methods Appl Fluoresc 4(2):022001
5. Berridge MJ (2016) The inositol trisphosphate/calcium signaling pathway in health and disease. Physiol Rev 96:1261–1296
6. Taylor CW, Konieczny V (2016) IP_3 receptors: take four IP_3 to open. Sci Signal 9:pe1
7. Yoshikawa F et al (1996) Mutational analysis of the ligand binding site of the inositol 1,4,5-trisphosphate receptor. J Biol Chem 271:18277–18284
8. Uchida K et al (2003) Critical regions for activation gating of the inositol 1,4,5-trisphosphate receptor. J Biol Chem 278:16551–16560
9. Bosanac I et al (2002) Structure of the inositol 1,4,5-trisphosphate receptor binding core in complex with its ligand. Nature 420:696–700
10. Seo M-D et al (2012) Structural and functional conservation of key domains in $InsP_3$ and ryanodine receptors. Nature 483:108–112
11. Rossi AM et al (2009) Synthetic partial agonists reveal key steps in IP_3 receptor activation. Nat Chem Biol 5:631–639
12. Fan G et al (2015) Gating machinery of $InsP_3R$ channels revealed by electron cryomicroscopy. Nature 527:336–341
13. Rossi AM et al (2010) Adenophostins: high-affinity agonists of IP_3 receptors. Curr Topics Membr 66:209–233
14. Ding Z et al (2010) Binding of inositol 1,4,5-trisphosphate (IP_3) and adenophostin A to the N-terminal region of the IP_3 receptor: thermodynamic analysis using fluorescence polarization with a novel IP_3 receptor ligand. Mol Pharmacol 77:995–1004
15. Wittmann HJ, Seifert R, Strasser A (2009) Contribution of binding enthalpy and entropy to affinity of antagonist and agonist binding at human and guinea pig histamine H_1-receptor. Mol Pharmacol 76:25–37
16. Cheng Y-C, Prusoff WH (1973) Relationship between the inhibition constant (K_I) and the concentration of inhibitor causing 50 per cent inhibition (IC_{50}) of an enzymatic reaction. Biochem Pharmacol 22:3099–3108
17. Nikolovska-Coleska Z et al (2004) Development and optimization of a binding assay for the XIAP BIR3 domain using fluorescence polarization. Anal Biochem 332(2):261–273

Chapter 10

Investigating the $InsP_3$ Receptor in Living Cells by Caged $InsP_3$

Xin Hui and Peter Lipp

Abstract

The inositol 1,4,5-trisphosphate receptor ($InsP_3R$) operates as an intracellular ligand-gated Ca^{2+} channel and plays a pivotal role in cellular Ca^{2+} homeostasis across all living cells. It is activated following membrane receptor–ligand interactions and stimulation of subsequent signaling cascades involving the enzymatic breakdown of the membrane lipid phosphatidyl-4,5-bisphosphate (PIP_2) into the membrane-delimited second messenger diacylglycerol (DAG) and the diffusible second messenger inositol-1,4,5-trisphophate ($InsP_3$). Modulation of $InsP_3R$'s activity is thus involved in a plethora of physiological and pathological processes. Here we combine membrane permeable photoactive caged-$InsP_3$ with Ca^{2+} imaging techniques in living cells to study the channel's in vivo properties. Using UV-flashes of variable energy, the activity properties of $InsP_3R$ can be investigated in great detail in its native environment.

Key words Ca^{2+}, $InsP_3$, $InsP_3R$, Photoactivation, UV-flash

1 Introduction

Intracellular Ca^{2+} works as a second messenger and directly modulates physiological and pathological processes in living cells. Consequently, the cytosol free Ca^{2+} concentration is fine-tuned in cellular homeostasis by multiple mechanisms [1, 2]. Among these, Ca^{2+}release from the endoplasmic reticulum (ER) into the cytosol through the ER membrane localized inositol 1,4,5-trisphosphate receptors ($InsP_3Rs$) is a major means to increase the cytosolic Ca^{2+} concentration rapidly [3, 4]. When inositol 1,4,5-trisphosphate ($InsP_3$), one product of the enzyme phospholipase C (PLC) that hydrolyzes membrane-delimited phosphatidylinositol 4,5-bisphosphate (PIP_2), binds to the $InsP_3R$, the receptor opens as an ion channel, allowing diffusion of Ca^{2+} from within the ER lumen (high Ca^{2+} concentration in the hundreds of micromolar range) to the cytosol displaying a resting concentration of around 100 nM Ca^{2+} [5]. Due to this steep concentration gradient Ca^{2+} ions can rapidly diffuse from the ER to the cytosol and thus cause

Gregory J. Miller (ed.), *Inositol Phosphates: Methods and Protocols*, Methods in Molecular Biology, vol. 2091,
https://doi.org/10.1007/978-1-0716-0167-9_10,

steep increases in the cytosolic Ca^{2+} concentration. After $InsP_3$ binding to $InsP_3R$, Ca^{2+} ions are accessible to the Ca^{2+} binding EF hands of $InsP_3R$ on its cytosolic part thus facilitating the opening of the ion gate. This indicates that $InsP_3$ is essential for the opening of $InsP_3R$ [6–8]. Interestingly, cytosolic Ca^{2+} is fostering the $InsP_3$ sensitivity of the receptor toward $InsP_3$ at lower Ca^{2+} concentrations, but, reportedly, high Ca^{2+} concentration can inhibit the channel opening, which indicates a complicated mechanism of $InsP_3R$ activity modulated by $InsP_3$ and Ca^{2+} [9–11]. $InsP_3R$'s channel activity can also be modulated by specific amino acid phosphorylation through various kinases and/or regulated by protein–protein interactions [12–15]. Because $InsP_3$ works as a signaling molecule in many signal pathways and its aforementioned complex relationship with Ca^{2+} [16], it is rather challenging to study the role of $InsP_3Rs$.

Here we describe a method utilizing membrane permeable photo-labile $InsP_3$ (ci-$InsP_3$) and UV-flash photolysis thereof to induce Ca^{2+} release from the ER in living cells. By simultaneous Ca^{2+} imaging and quantitative analysis of the resulting Ca^{2+} transients we were able to study $InsP_3R$'s function in living cells in great details. We will use the recently revealed interaction of the $InsP_3R$ with the β3 subunit of the voltage-gated Ca^{2+} channel as a use-case [15].

2 Materials

2.1 Reagents

1. Membrane permeable caged $InsP_3$ (ci-$InsP_3$/PM, SiChem) (*see* **Note 1**).
2. Membrane permeable Ca^{2+} indicator (Fluo-4 AM, Thermo Fisher) (*see* **Note 2**).
3. DMSO (Sigma-Aldrich).
4. Pluronic F-127 (Thermo Fisher).
5. 12-well cell culture plate (Thermo Fisher).
6. 20 mm-diameter glass coverslip (Thermo Fisher).

2.2 Main Equipment

1. Nikon upright microscopy (TE-2000E, Nikon).
2. Fast 2D-kilobeam array scanner (Infinity-3, VisiTech).
3. EMCCD-camera (iXon 887, Andor).
4. VoxCellScan software (Visitech International).
5. Oil immersion objective with high UV transmission (40×, NA 1.3, S-Fluor, Nikon).
6. Diode-pumped solid-state laser at 491 nm (Calypso, Cobolt).
7. Xenon flash lamp system (JML-C2, Rapp OptoElectronic).

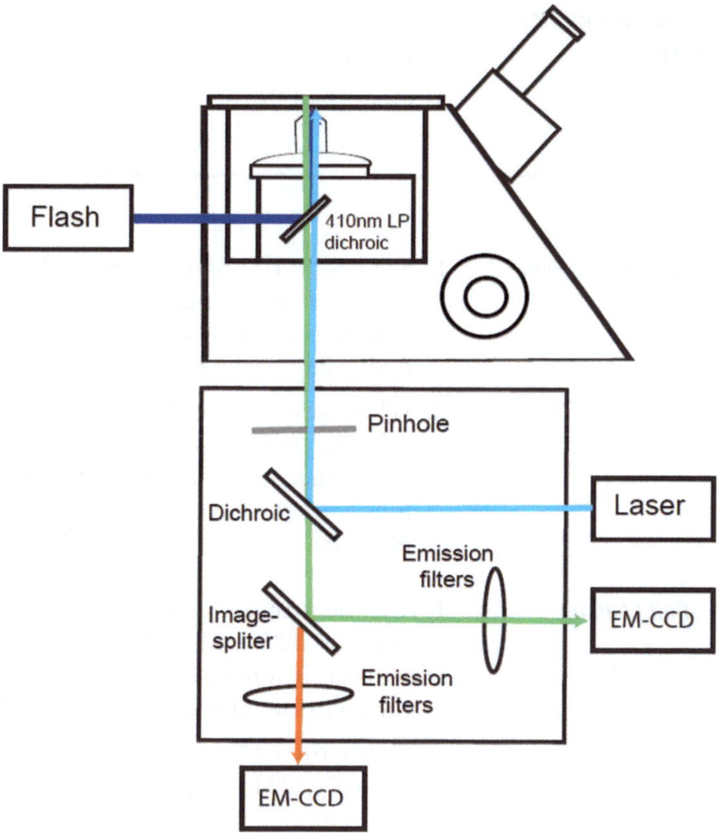

Fig. 1 Schematic illustration of fast-scanning confocal system based on an inverted microscope. UV-flash light was reflected to the specimens by a 410 nm long-pass dichroic mirror. The Ca^{2+} imaging was performed by a fast-scanning confocal system with 491 nm laser and respective optical mirrors

8. Quartz light guide (Rapp OptoElectronic) (*see* **Note 3**).
9. Dichroic mirror (410 nm LP, AHF Analysentechnik).

Photo-activation of ci-$InsP_3$ and simultaneous Ca^{2+} imaging in living cells were performed on a fast-scanning confocal system, using an inverted microscope (TE-2000E, Nikon) with an oil immersion objective (40×, NA 1.3 S-Fluor, Nikon). The microscope was attached to a fast 2D-kilobeam array scanner (Infinity-3; VisiTech Int.) that simultaneously scans 2500 parallel laser beams across the specimen and projects the resulting fluorescence images on two spectrally separated EMCCD-cameras (iXon 887, Andor Technologi) with proper image dichroic and filters (Fig. 1). A solid-state laser at 491 nm (Calypso , Cobolt) was used for Fluo-4 excitation. The entire setup was integrated and controlled by Vox-CellScan software (VisiTech Int., UK) (*see* **Note 4**).

2.3 Software for Data Analysis

1. NIH ImageJ.
2. Wavemerics Igor Pro.
3. GraphPad Prism.

3 Method

3.1 Cell Preparation

1. Seed 5 × 10^4 mammalian cells (e.g., HEK293 cells or COS cells) on 20 mm-diameter glass coverslip within 12-well cell culture plate with 1 mL complete growth medium 24 h before Ca^{2+} imaging (*see* **Note 5**).
2. When transfection is necessary, 12–18 h after cell seeding, suitable transfection reagents such as Lipofectamine 2000 or NanoJuice, could be applied on these cells in their 12-well plate following the manufactory's protocol with desired plasmids (*see* **Note 6**).

3.2 Cell Loading

1. Tyrode's buffer is used during the imaging process, which contains 135 mM NaCl, 5.4 mM KCl, 2 mM $MgCl_2$, 10 mM glucose, 1.8 mM $CaCl_2$, and 10 mM HEPES. The pH value of the solution adjusts to 7.35 with NaOH.
2. Dilute the Ca^{2+} indicator Fluo-4 AM to 1 μM and caged $InsP_3$, ci-$InsP_3$/PM to 5 μM with Tyrode's buffer as working solution (*see* **Note 7**).
3. Transfer the coverslip with the cells from the 12-well cell plate to a custom-made chamber and incubate these cells in 1 mL of the above working solution for 30 min at room temperature.
4. Wash the cells with normal Tyrode's buffer carefully and incubate in normal Tyrode's buffer for 10 min as de-esterification (*see* **Note 8**).

3.3 Ca^{2+} Imaging and $InsP_3$ Photoactivation

1. Mount the chamber on to the microscope with a 40×, NA 1.3, S-Fluor, oil immersion objective. Fluo-4 is excited with the 491 nm laser and fluorescence emission collects through a 510 nm long-pass filter. Images (120 × 128 pixels) are recorded with EMCCD-cameras at 5 frames per second with 4 × 4 binning (Fig. 2) (*see* **Note 9**).
2. Photoactivation of ci-$InsP_3$ was achieved by introducing a bright light flash from a Xenon flash lamp system. Desired spectrum (360–400 nm) of the flash was delivered to the specimen by reflection as a 410 nm long-pass dichroic mirror (Fig. 1). The energy contained in a single flash could be adjusted by choosing the appropriate lamp capacitance (1000 μF, 2000 μF or 3000 μF) and/or changing the charge voltage continuously between 0 and 385 V. During the Ca^{2+} imaging experiment, the VoxCellScan software controlled the

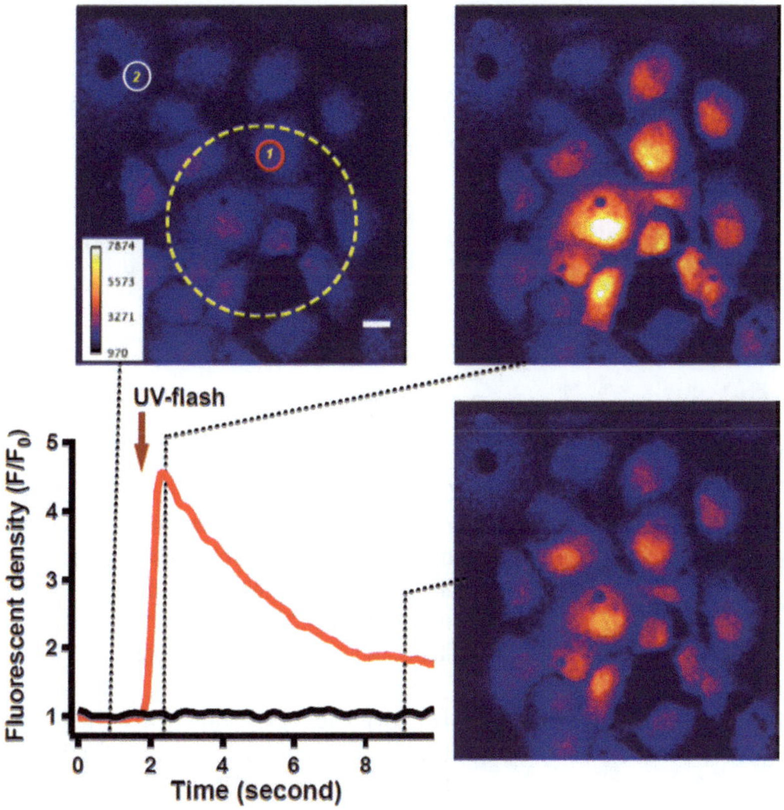

Fig. 2 Photoactivation of ci-$InsP_3$ in living HEK cells. The color-coded images show the ci-$InsP_3$ and Fluo-4 loaded cells before and after UV-flash at the indicated time. The fluorescent changes at ROI1 and ROI2 labeling in the upper left image were plotted with time as the red curve for ROI1 and black curve for ROI2 in the lower left panel. The yellow dash ring in the upper left image indicates the center of the illuminated region. Scale bar is 10 μm

triggering of the UV flash at certain time point(s) and the resulting fluorescence images were save as a stack of tiff-images. Due to experimental issues described below each field-of-view was only illuminated once (*see* **Note 10**).

3.4 Image Analysis

1. Open the saved file with the software ImageJ and check the image quality. Color-coded 2D images of Ca^{2+} signal were reconstructed in ImageJ from original 16-bit grey scale images by applying appropriate look-up table (Fig. 2) (*see* **Note 11**).
2. Choose the regions of interest (ROI) and quantify the fluorescent intensity over time of the series of images in the ROIs by appropriate functions in ImageJ (Fig. 2). Here we marked a cell within the UV-flash region with ROI1 and while ROI2 highlighted a cell outside of the UV-illumination (*see* **Note 10**). The resulting series of intensities were saved as a "txt" file.

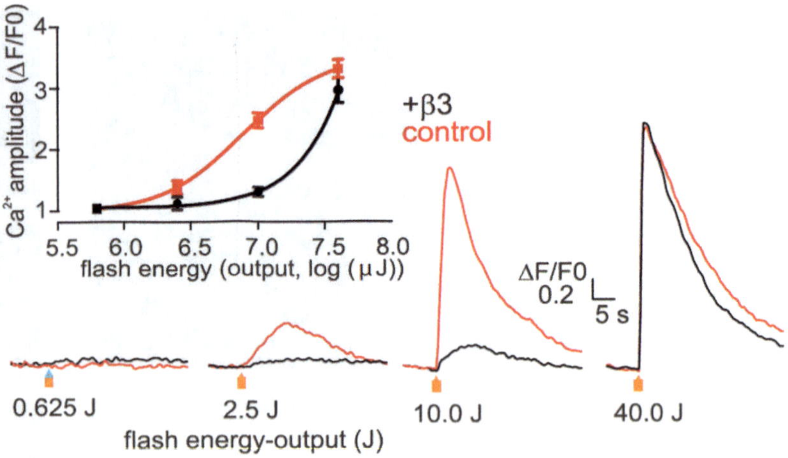

Fig. 3 Modulation of the $InsP_3R$ by the ß3 subunit of the voltage operated Ca^{2+} channel. The bottom row shows the Fluo-4 signals in cells expressing the ß3-subunit (black) and control cells (red) following UV-flash (arrow indicated) stimulation with a serial of output energy. The upper panel shows a dose-dependent relationship between the energy outputs of UV-flash and Ca^{2+} amplitude. (Adapted from Cell Reports 22:1339–1349)

3. Import the "txt" files into the software Igor Pro to plot the traces. In order to account for inhomogeneous Fluo-4 distribution between cells, self-ratio F/F_0 was calculated, whereas F_0 is the average of the resting fluorescence intensity at the first 5–10 points (Fig. 2). Quantitatively measure the peak of Ca^{2+} signal by Igor Pro (*see* **Note 12**).
4. According to the formula, $E = ½ \times C \times U^2$, whereas E is the output energy of single flash, C is the capacitance, and U is the voltage of the capacitors, output energy of each single flash can be calculated with a given voltage and capacitance. Applying a series of UV-flashes with variable flash energy, the respective Ca^{2+} signals can be recorded and further quantified (Fig. 3) by following Method 3.3 (*see* **Note 13**).
5. We plotted the UV-flash energy/Ca^{2+} amplitude relationship using the software Prism (GraphPad Software) and fitted the data points with a dose-dependent curve. In this use-case (Fig. 3), we compared the resulting dose-dependent curves of β3 (subunit of voltage-dependent $Ca2^+$ channel) overexpression cells (black) and control cells (red), $InsP_3Rs$ in control cells is much more sensitive to $InsP_3$ than those in β3 overexpression cells (*see* **Note 14**).

4 Notes

1. The membrane permeable caged $InsP_3$, ci-$InsP_3$/PM, should be stored at −20 °C as powder and dissolved in dry DMSO at 1 mM just before use, avoiding degradation.
2. The membrane permeable Ca^{2+} indicator, Fluo-4 AM should be dissolve in Pluronic F-127/dry DMSO at 1 mM as a stock solution and store at −20 °C.
3. All optical instruments and parts therein used for UV-flash should display very good UV transmission for efficient photo-active experiments. This holds particularly true for the microscope objective used.
4. Epifluorescence microscopes are also suitable for Ca^{2+} imaging in living cells as long as no high spatial information is required. Using an image splitter setup, more than one channel of fluorescent can be acquired simultaneously. Alternatively, using a single camera, emission filters have to be changed between alternating images.
5. Cell culture parameters were optimized according to the proliferation of the cells and the culture time before the Ca^{2+} assay. Around 70% confluency of cells on the coverslips appeared optimal for ci-$InsP_3$ photoactivation and Ca^{2+} imaging. After seeding of the cells, they should be given sufficient time for settling and adhering, in general 24 h is adequate.
6. If necessary, transfection protocols should be optimized to the individual cell types and plasmids used.
7. For all the operations with caged $InsP_3$ any exposure to light should be strictly avoided. We perform all handling and experiments in dim red light. The working concentration of caged $InsP_3$ may have to be adjusted to the cell type used in order to ensure sufficient loading of caged $InsP_3$ into the cells.
8. Please carefully exchange the solution in the cell chamber to avoid washing away of adherent cells. With extended loading times or higher loading temperatures, Fluo-4 will accumulate in the cell's nucleus.
9. In order to obtain a high dynamic range of the fluorescent signal several parameters need optimization. Firstly, the output of the excitation lasers. Higher laser power yields high fluorescence signals but also cause stronger bleaching and induce laser toxicity for the cells. Secondly, increasing the binning of the camera's pixels can enhance the signal but sacrifices spatial resolution. Thirdly, the exposure time should also be optimized according to the required temporal resolution. These three points are the major considerations for Ca^{2+} imaging and might need particular optimization prior to the studies.

10. According to the fluorescence signals from Fluo-4, the output range of the UV-flashes requires adjustment and optimization and should be set between causing almost no Ca^{2+} increase to the maximum Ca^{2+} transient in the cells. Due to the optical property of the light guide and the objective lenses, the illuminated region of UV-flash may be smaller than the field of view seen by the camera. Thus, the field-of-view ought to be determined in additional control experiments, for example by using supramaximal UV-flashes. Because of the complex relationship between $Ca^{2+}/InsP_3$ and the $InsP_3R$ and its multiple modulation by phosphorylation we only investigated naïve cells.
11. Briefly check the quality of images, excluding some images with focus shift, image movement, unexpected and disturbing events, and so on. Also make sure that the fluorescent signal stays within the dynamic range of the camera, avoiding any signal saturation in individual pixels.
12. The background signal from the camera should be subtracted before F/F_0 calibration. Multiple indexes could be used to quantify Ca^{2+} mobilization, for instants, the peak of amplitude, the area under the curve, the speed of upstroke, the decay of the amplitude according to different interests. Here we used the peak of the amplitude from Ca^{2+} signal as a major index to indicate the activity of $InsP_3R$.
13. Due to differences in the transmission of optical parts in the particular set-up used, the energy output from the UV-flash lamp will need careful adjustment. It needs to note that the output energy of UV-flash is a theoretical value from the UV lamp. The practical value of UV energy putting on the cells requires restricted measurement.
14. With lower energies of the UV-flash we found the upstroke of the resulting Ca^{2+} transients to be substantially slowed down, highlighting the low positive feedback in the system with small cytosolic Ca^{2+} concentrations and low $InsP_3$ (lower row in Fig. 3).

References

1. Berridge MJ, Lipp P, Bootman MD (2000) The versatility and universality of calcium signalling. Nat Rev Mol Cell Biol 1:11–21. https://doi.org/10.1038/35036035
2. Berridge MJ, Bootman MD, Roderick HL (2003) Calcium signalling: dynamics, homeostasis and remodelling. Nat Rev Mol Cell Biol 4:517–529. https://doi.org/10.1038/nrm1155
3. Bootman MD (2012) Calcium signaling. Cold Spring Harb Perspect Biol 4:a011171. https://doi.org/10.1101/cshperspect.a011171
4. Berridge MJ (2012) Calcium signalling remodelling and disease. Biochem Soc Trans 40:297–309. https://doi.org/10.1042/BST20110766
5. Foskett JK, White C, Cheung KH, Mak DOD (2007) Inositol trisphosphate receptor Ca^{2+} release channels. Physiol Rev 87:593–658.

https://doi.org/10.1152/physrev.00035.2006

6. Taylor CW, Tovey SC (2010) IP3 receptors: toward understanding their activation. Cold Spring Harb Perspect Biol 2: a004010–a004010. https://doi.org/10.1101/cshperspect.a004010
7. Fan G, Baker ML, Wang Z et al (2015) Gating machinery of InsP3R channels revealed by electron cryomicroscopy. Nature 527:336–341. https://doi.org/10.1038/nature15249
8. Várnai P, Balla A, Hunyady L, Balla T (2005) Targeted expression of the inositol 1,4,5-triphosphate receptor (IP3R) ligand-binding domain releases Ca^{2+} via endogenous IP3R channels. Proc Natl Acad Sci U S A 102:7859–7864. https://doi.org/10.1073/pnas.0407535102
9. Mak D, McBride S, Foskett JK (1999) Inositol 1,4,5-tris-phosphate activation of inositol tris-phosphate receptor Ca^{2+} channel by ligand tuning of Ca^{2+} inhibition (vol 95, pg 15821, 1998). Proc Natl Acad Sci U S A 96:3330–3330
10. Bootman MD, Lipp P (1999) Ringing changes to the 'bell-shaped curve'. Curr Biol 9: R876–R878
11. Taylor CW, da Fonseca PCA, Morris EP (2004) IP3 receptors: the search for structure. Trends Biochem Sci 29:210–219. https://doi.org/10.1016/j.tibs.2004.02.010
12. Szado T, Vanderheyden V, Parys JB et al (2008) Phosphorylation of inositol 1,4,5-trisphosphate receptors by protein kinase B/Akt inhibits Ca^{2+} release and apoptosis. Proc Natl Acad Sci U S A 105:2427–2432. https://doi.org/10.1073/pnas.0711324105
13. Patterson RL, Van Rossum DB, Barrow RK, Snyder SH (2004) RACK1 binds to inositol 1, 4, 5-trisphosphate receptors and mediates Ca^{2+} release. Proc Natl Acad Sci U S A 101:2328
14. Hui X, Reither G, Kaestner L, Lipp P (2014) Targeted activation of conventional and novel protein kinases C through differential translocation patterns. Mol Cell Biol 34:2370–2381. https://doi.org/10.1128/MCB.00040-14
15. Belkacemi A, Hui X, Wardas B et al (2018) IP3 receptor-dependent cytoplasmic Ca^{2+} signals are tightly controlled by Cavβ3. Cell Rep 22:1339–1349. https://doi.org/10.1016/j.celrep.2018.01.010
16. Smrcka AV, Brown JH, Holz GG (2012) Role of phospholipase Cε in physiological phosphoinositide signaling networks. Cell Signal 24:1333–1343. https://doi.org/10.1016/j.cellsig.2012.01.009

Chapter 11

Optical Measurement of Synaptic Vesicle Fusion and Its Inhibition by Inositol Pyrophosphate in Primary Cultured Hippocampal Neurons

Sung Hyun Kim

Abstract

Optical molecular and cellular imaging has provided considerable information in various life science research fields. In particular, pHluorin-based assays allow us to specifically investigate presynaptic physiological processes, such as synaptic transmission and synaptic retrieval. Here, I describe quantitative synaptic transmission in primary cultured hippocampal neurons using a vGlut-pHluorin system combined with a high-fidelity optical imaging system and its regulation by inositol pyrophosphate.

Key words Synaptic transmission, Synaptic vesicle, vGlut-pHluorin (vG-pH), Inositol pyrophosphate, IP6K

1 Introduction

Many aspects of cellular function, including mitochondria motility [1], plasma membrane dynamics [2] and Ca^{2+} dynamics (influx and clearance) [3], among others, can be investigated using optical live-imaging systems. One of the most important applications in the neuroscience field is synaptic activity monitoring. Optical imaging systems combined with specific genetic and chemical indicators (e.g., FM dyes, synaptopHluorin) have recently enabled the investigation of presynaptic activity. One such system is the pHluorin-based assay system, which allows direct monitoring of presynaptic physiology (synaptic vesicle fusion and synaptic vesicle retrieval). pHluorin is a mutant form of GFP (green fluorescent protein) whose fluorescence is quenched upon protonation because of its pKa value (~7.1) [4, 5]. Fusion of pHluorin to one of the intraluminal loops of a synaptic vesicular membrane protein, in this case, vesicular glutamate transporter1 (vGlut1), provides high efficiency measurements of synaptic vesicle fusion because of the very low surface expression of vGlut1 under normal conditions [6, 7]. When

Gregory J. Miller (ed.), *Inositol Phosphates: Methods and Protocols*, Methods in Molecular Biology, vol. 2091,
https://doi.org/10.1007/978-1-0716-0167-9_11,

synaptic vesicle fusion is evoked by stimulation, vGlut1-pHluorin (vG-pH) is exposed to the extracellular buffer, which is pH 7.4, causing alkalization and dequenching of vG-pH, resulting in a 20–25-fold increase in its brightness compared with acidic resting conditions. After endocytosis of synaptic vesicles, vG-pH is instantaneously quenched by reacidification via a proton pump on the synaptic vesicle membrane. In this chapter, I describe the monitoring of synaptic vesicle fusion by measuring vG-pH responses to 100 action potential stimuli at 10 Hz through 50–100 synaptic boutons from a single neuron. Fluorescence signals at individual boutons are normalized to the intensity value obtained by rapid alkalization of the entire labeled synaptic vesicle pool using NH_4Cl [8, 9].

2 Materials

2.1 Primary Culture of Hippocampal Neurons

1. Rat: 0- to 1-day-old Sprague-Dawley rats.
2. Hippocampus: Dissected hippocampus containing the CA1-CA3 region, but not dentate gyrus.
3. Cultureware: Culture dish (35 mm), coverslip (22 × 22 mm).
4. Coating buffer: Poly-L-ornithine (10 μg/ml) in sterile filtered water.
5. Digestion buffer: Trypsin (10 mg/ml) and DNase I (0.5 mg/ml) in digestion buffer (137 mM NaCl, 5 mM KCl, 7 mM Na_2HPO_4, 25 mM HEPES, pH 7.2).
6. Dissociation buffer: HEPES (25 mM) and $MgSO_4$ (12 mM).
7. Hank's Balanced Salt Solution (HBSS): 350 mg $NaHCO_3$, 260 mg HEPES (pH 7.2) per bottle (1 L).
8. Culture medium: Neurobasal medium (500 ml) containing 5 ml of GlutaMAX and 10 ml of B-27 supplement.

2.2 Transfection

1. 2 mM $CaCl_2$.
2. 2× HeBS (273 mM NaCl, 0.5 mM KCl, 1.4 mM Na_2HPO_4, 15 mM D-glucose, and 42 mM HEPES, pH 7.1),
3. Distilled water.
4. Plasmid: vGlut1-pH (vG-pH), shRNA targeting IP6K (inositol hexakisphosphate kinase).

2.3 Optical Imaging System

1. Fluorescence microscope with 40× (~1.3 NA) Fluor objective lens.
2. Filter set (excitation 450–490 nm, emission 500–550 nm and 498 nm dichroic filters).

3. EMCCD camera (electron-multiplying charge-coupled device).
4. Diode pumped 488 nm laser.
5. Isolated pulse generator.
6. Imaging buffer: Tyrode's buffer (119 mM NaCl, 2.5 mM KCl, 2 mM $CaCl_2$, 2 mM $MgCl_2$, 25 mM HEPES, 30 mM glucose, 10 μM 6-cyano-7-nitroquinoxaline-2,3-dione, and 50 μM D,L-2-amino-5-phosphonovaleric acid), adjusted to pH 7.4.

2.4 Image Analysis Programs

1. ImageJ with the Time-series analyzer plugin.
2. Origin Pro-8.0.
3. Microsoft Excel.

3 Methods

3.1 Primary Culture of Rat Hippocampal Neurons

1. Carefully dissect the hippocampal CA3–CA1 region from postnatal 0- to 1-day-old Sprague-Dawley rats [10].
2. Incubate dissected regions with 2 ml of digestion buffer for 10 min at 37 °C.
3. Dissociate tissues in 2 ml of dissociation buffer by pipetting gently 20–25 times.
4. Centrifuge dissociated cells at 340 × *g* for 10 min at 7 °C.
5. Remove supernatant and resuspend cells in culture medium.
6. Count cells using a hemocytometer.
7. Plate the appropriate number of cells (4.5×10^4) onto poly-ornithine–coated cover glass in culture medium.
8. Incubate neurons at 37 °C in a humidified 5% CO_2 atmosphere.

3.2 Transfection

3.2.1 All Procedures for Transfection Are Carried Out on a Clean Bench at Room Temperature

1. After culturing for 8 days in vitro (8 DIV), replace 80% of the culture medium in the dish with plain Neurobasal medium.
2. Place 9 μg of plasmid with/without shRNA targeting IP6K into a microcentrifuge tube.
3. Add distilled water to a total volume of 40.5 μl and mix gently.
4. Add 4.5 μl of 2 mM $CaCl_2$ and mix well.
5. Add 45 μl of 2× HeBS and mix vigorously.
6. Apply 12 μl of mixture to cells in the dish.
7. Incubate for 10–15 min in a CO_2 incubator.
8. Replace transfection mixture with culture medium.
9. Incubate cells for an additional 6–13 days in a CO_2 incubator.

3.3 Optical Imaging and Analysis

3.3.1 Neurons at 14–21DIV Are Utilized for Optical Live-Cell Imaging. All Live-Cell Imaging Experiments Are Performed at 30 °C

1. Mount cover glass-plated neurons in the stimulation chamber and perfuse with a saline-based buffer (Tyrode's buffer).
2. To identify vG-pH with/without shRNA-targeting IP6K transfected cells, briefly stimulate neurons with hexa-pulse action potentials (6 APs) at 30 Hz. APs are evoked by passing a 1-ms current pulse through platinum-iridium electrodes from an isolated current stimulator.
3. Once neurons expressing vG-pH with/without shRNA targeting IP6K are found, stimulate cells with 100 AP at 10 Hz, followed by application of NH_4Cl. For time-lapse recordings, acquire images with an electron-multiplying CCD camera, collecting 400 frames at 2 Hz with a 50-ms exposure.
4. Analyze time-lapse images with ImageJ using the Time-series analyzer plugin. Select vG-pH positive-responsive synaptic boutons as a region of interest (ROI), and measure intensities of vG-pH on synaptic boutons upon stimulation.
5. Analyze fluorescence traces using Origin Pro-8.0. After subtracting background signal, normalize traces to the peak signal of NH_4Cl.
6. Calculate the percentage (%) of vG-pH peak signal after 100 AP stimuli by normalizing to the maximum signal of NH_4Cl in neurons with/without shRNA targeting IP6K.

Acknowledgments

This work was supported by National Research Foundation of Korea Grant 2017R1A2B4007019 funded by the Ministry of Science and Information and Communication Technology.

References

1. Bae JR, Kim SH (2016) Impairment of SOD1-G93A motility is linked to mitochondrial movement in axons of hippocampal neurons. Arch Pharm Res 39(8):1144–1150. https://doi.org/10.1007/s12272-016-0798-5
2. Kim DJ, Kim SH, Lim CS, Choi KY, Park CS, Sung BH, Yeo MG, Chang S, Kim JK, Song WK (2006) Interaction of SPIN90 with the Arp2/3 complex mediates lamellipodia and actin comet tail formation. J Biol Chem 281 (1):617–625
3. Kim SH, Ryan TA (2013) Balance of calcineurin Aα and CDK5 activities sets release probability at nerve terminals. J Neurosci 33 (21):8937–8950
4. Sankaranarayanan S, De Angelis D, Rothman JE, Ryan TA (2000) The use of pHluorins for optical measurements of presynaptic activity. Biophys J 79(4):2199–2208
5. Miesenbock G, De Angelis DA, Rothman JE (1998) Visualizing secretion and synaptic transmission with pH-sensitive green fluorescent proteins. Nature 394(6689):192–195. https://doi.org/10.1038/28190
6. Voglmaier SM, Kam K, Yang H, Fortin DL, Hua Z, Nicoll RA, Edwards RH (2006) Distinct endocytic pathways control the rate and extent of synaptic vesicle protein recycling. Neuron 51(1):71–84. https://doi.org/10.1016/j.neuron.2006.05.027
7. Balaji J, Ryan TA (2007) Single-vesicle imaging reveals that synaptic vesicle exocytosis and endocytosis are coupled by a single stochastic mode. Proc Natl Acad Sci U S A 104 (51):20576–20581

8. Lee TS, Lee JY, Kyung JW, Yang Y, Park SJ, Lee S, Pavlovic I, Kong B, Jho YS, Jessen HJ, Kweon DH, Shin YK, Kim SH, Yoon TY, Kim S (2016) Inositol pyrophosphates inhibit synaptotagmin-dependent exocytosis. Proc Natl Acad Sci U S A 113(29):8314–8319. https://doi.org/10.1073/pnas.1521600113
9. Kim SH, Ryan TA (2010) CDK5 serves as a major control point in neurotransmitter release. Neuron 67(5):797–809
10. Kyung JW, Cho IH, Lee S, Song WK, Ryan TA, Hoppa MB, Kim SH (2017) Adaptor Protein 2 (AP-2) complex is essential for functional axogenesis in hippocampal neurons. Sci Rep 7:41620. https://doi.org/10.1038/srep41620

Chapter 12

Competitive Fluorescent Ligand Assay for Inositol 1,4,5-Trisphosphate

Akihiko Tanimura and Satoshi Shuto

Abstract

We present a novel method, termed competitive fluorescent ligand assay for inositol 1,4,5-trisphosphate (CFLA-IP_3), to measure inositol 1,4,5-trisphosphate (IP_3). This method is based on fluorescence resonance energy transfer (FRET) between two fluorescent molecules, a fluorescent IP_3-binding protein and its fluorescent ligand. Binding of these fluorescent molecules generates a FRET signal, and the IP_3-dependent decrease in the FRET signal due to displacement of the fluorescent ligand is detected by fluorescence microscopy.

Key words Inositol 1,4,5-trisphosphate, IP_3 receptor, Ligand binding domain, Fluorescent ligand, Fluorescence resonance energy transfer, Fluorescence microscopy

1 Introduction

Inositol 1,4,5-trisphosphate (IP_3) is an important intracellular messenger produced by phospholipase C-dependent hydrolysis of phosphatidylinositol 4,5-bisphosphate and releases Ca^{2+} from intracellular stores via IP_3 receptors (IP_3Rs) [1] IP_3 functions as an important second messenger in G-protein-coupled receptor- and receptor tyrosine kinase-mediated pathways. The production of IP_3 is traditionally measured by a radioactive phosphatidylinositol turnover assay and anion-exchange chromatography [2–4]. However, this method is time-consuming, laborious, and expensive.

Recently, two methods have become available to measure IP_3 concentrations, namely, AlphaScreen technology, which monitors singlet oxygen-induced chemiluminescence of thioxene compounds, and fluorescence polarization to detect the binding of IP_3-binding proteins and fluorescent or biotinylated analogues of IP_3 [5, 6]. Although these methods do not require separation of IP_3, the need for large amounts of purified IP_3-binding proteins

Gregory J. Miller (ed.), *Inositol Phosphates: Methods and Protocols*, Methods in Molecular Biology, vol. 2091,
https://doi.org/10.1007/978-1-0716-0167-9_12,

and accurate quantification of the concentrations of functional binding sites limits their applicability.

To overcome this difficulty, we developed a new method termed competitive fluorescent ligand assay for inositol 1,4,5-trisphosphate (CFLA-IP_3). In this method, fluorescence resonance energy transfer (FRET) upon the binding of a fluorescent ligand (FLL) to a fluorescent IP_3 sensor protein, the CFP-coupled ligand-binding domain of IP_3R, was determined by fluorescence microscopy. The competitive binding of IP_3 and FLL to the fluorescent IP_3 sensor protein caused an IP_3-dependent decrease in the FRET signal due to the displacement of FLL. This method allowed us to estimate the cytosolic concentrations of IP_3 in ATP-stimulated and unstimulated cells, and to evaluate IP_3R analogues [7].

2 Materials

Prepare all solutions using ultrapure water and analytical grade reagents.

2.1 IP_3 Sensor Plasmids

A fluorescent IP_3 sensor consisting of a HisTag, CFP (Cerulean), and the ligand-binding domain of type 2 IP_3R (Fig. 1) was constructed by conventional molecular cloning techniques [7].

2.2 Preparation of IP_3 Sensor Beads

1. Culture COS-7 cells in a 100 mm dish and transfect them with 15 μg/mL of the plasmid harboring the fluorescent IP_3 sensor gene using Lipofectamine 2000 (Invitrogen, Carlsbad, CA, USA) according to the manufacturer's instructions.
2. Two days after transfection, detach cells expressing the fluorescent IP_3 sensor protein from the culture dish using Ca^{2+}-free phosphate-buffered saline (PBS) (containing 1 mM EDTA).
3. Wash the cells with HisTALON™ Equilibration Buffer (EQB) and permeabilize them with EQB containing 200 μg/mL saponin for 15–30 min at 4 °C with continuous agitation.
4. Remove the supernatant containing the fluorescent IP_3 sensor protein and mix it with TALON beads (25–50 μL of 50% slurry) for 30–60 min at 4 °C.
5. Wash the fluorescent IP_3 sensor-bound beads (IP_3 sensor beads) with EQB and keep them at 4 °C until use (*see* **Note 1**).

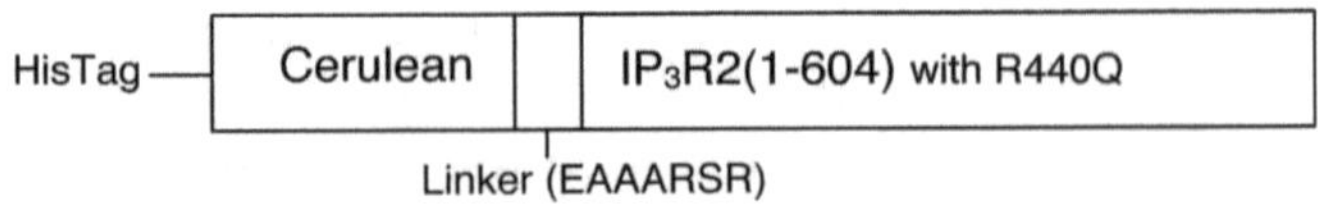

Fig. 1 Structure of the fluorescent IP_3 sensor. This sensor consists of a HisTag, CFP (Cerulean), and the ligand-binding domain of the rat type2 IP_3R. The ligand-binding domain harbors the R440Q mutation

2.3 Experimental Chamber

1. Prepare experimental chambers with a plastic cylinder (7 mm in diameter and 7 mm in height) glued to a round glass coverslip (25 mm in diameter) using silicone rubber adhesive. Alternatively, glue the plastic cylinders to a glass-bottom dish (Iwaki, Shizuoka, Japan).
2. Place 10–20 μL of 1.85 mg/mL Cell-Tak solution or 1 mg/mL poly-L-lysine solution on the top of the experimental chamber and wait 1–2 h.
3. Wash the experimental chamber with distilled water and add IP_3 sensor beads (*see* **Note 2**).
4. Immobilize the beads to the bottom of experimental chamber by the centrifugation (500 rpm, 40 s).

2.4 Fluorescent Low-Affinity Ligand (FLL)

FLL was synthesized by introducing fluorescein into the 5′-position of an adenophostin A analog that lacks adenine at the ribose anomeric position [7].

3 Methods

Carry out all procedures at room temperature unless otherwise specified.

3.1 Preparation of Samples

1. Prepare 60 nM (or 200 nM) FLL using intracellular-like medium containing 125 mM KCl, 19 mM NaCl, and 10 mM HEPES-KOH (pH 7.3).
2. Prepare appropriate concentrations of IP_3 solutions and sample solutions using the same buffer.
3. Mix the sample solution and the FLL solution at a ratio of 1:1.

3.2 Acquisition of Fluorescence Images

1. Place the experimental chamber with immobilized IP_3 sensor beads on the fluorescence microscope.
2. Consecutively capture fluorescence images of IP_3 sensor beads (5–20 s intervals) with an excitation wavelength of 425 nm and dual emission wavelengths of 480 and 535 nm (Fig. 2, *see* **Note 3**).
3. Apply IP_3 standards and samples at an interval of 3–5 min during the capture of fluorescence images.

3.3 Image Analyses to Determine Net Changes in the Fluorescence of IP_3 Sensor Beads

1. Analyze the fluorescence intensity in the area containing (Area A) and lacking (Area B) IP_3 sensor beads using image processing software (Fig. 2A, B).
2. Subtract the fluorescence intensity in Area B from that in Area A (Fig. 2C).

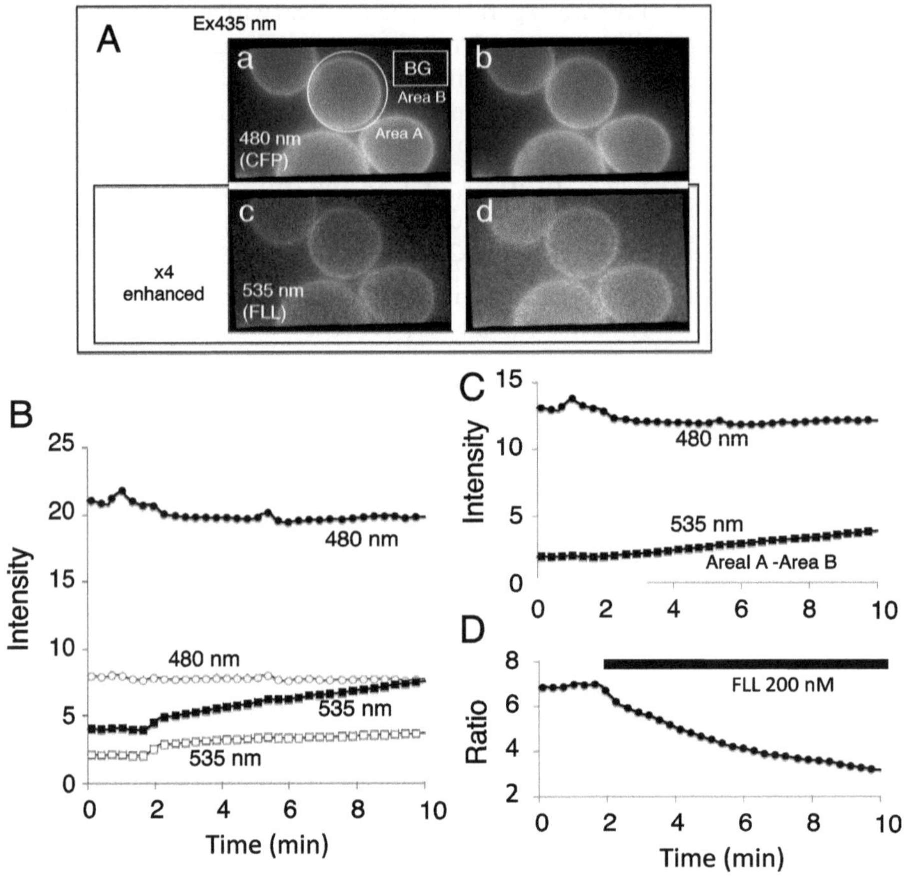

Fig. 2 Image analyses to determine the FLL-induced FRET signal. (**A**) Fluorescence images of the IP_3 sensor beads before (a and c) and after (b and d) the application of FLL (200 nM) acquired with an excitation wavelength of 425 nm and dual emission wavelengths of 480 nm (a and b) and 535 nm (c and d). (**B**) Changes in the fluorescence intensity in Area A (IP_3 sensor beads with background) and Area B (background). Upon addition of FLL, fluorescence at 535 nm (open symbols) in Area B increased rapidly due to direct excitation. In Area A, on the other hand, after the rapid increase upon addition of FLL, fluorescence at 480 and 535 nm slowly decreased and increased, respectively, due to FRET caused by binding of FLL to the fluorescent proteins on the beads. (**C**) To exclude signals arising from direct excitation of FLL, the fluorescence intensity in Area B was subtracted from that in Area A. (**D**) The background-subtracted fluorescence intensities (**c**) were used to determine the FLL-induced changes in the fluorescence emission ratio (480/535 nm)

3. Calculate the fluorescence emission ratio (480/525 nm) at different concentrations of IP_3 and samples with a fixed concentration of FLL using the net changes in fluorescence intensities obtained in **step 2** (Fig. 2D, *see* **Note 4**).

3.4 Determination of IP_3 Concentrations

Plot the fluorescence ratio (480/535 nm) of IP_3 sensor beads against the IP_3 concentrations using a computer spreadsheet programmed with graphical output. A line graph should be generated by joining the points (Fig. 3d). The IP_3 concentration in the sample

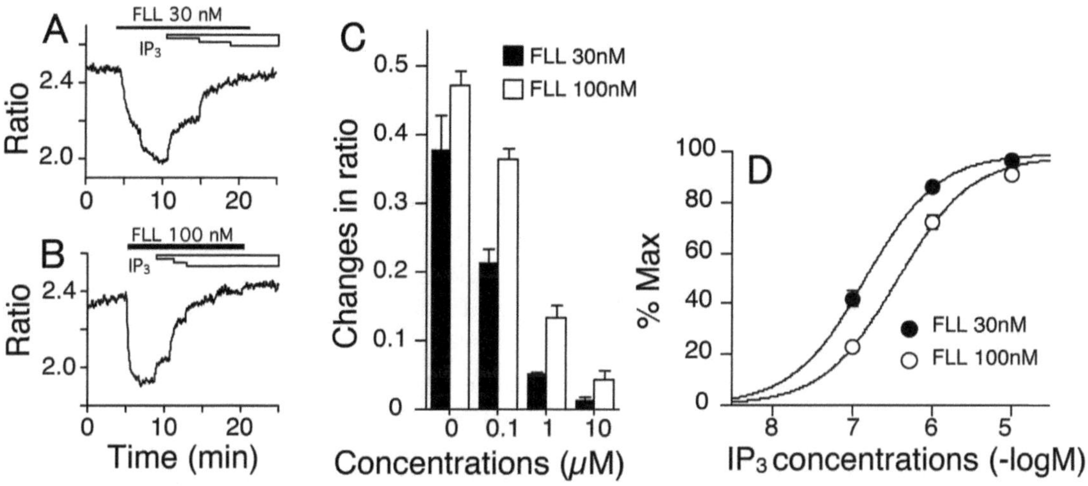

Fig. 3 Effects of FLL and subsequent addition of IP_3 on the fluorescence ratio of the IP_3 sensor beads. (**a**) Changes in the fluorescence ratio with 30 nM FLL and 0.1, 1, or 10 μM IP_3. (**b**) Changes in the fluorescence ratio with 100 nM FLL and 0.1, 1, or 10 μM IP_3. (**c**) Summarized results of (**a**) and (**b**). (**d**) Concentration-response curve of IP_3 and FRET signal recovery in the presence of 30 and 100 nM FLL. The IC_{50} values of IP_3 with 30 and 100 nM FLL were 70 and 250 nM, respectively

solution can be read off the graph. Alternatively, the IP_3 concentration can be calculated from the equations provided below.

3.5 Determination of IP_3 Concentrations Using Equations

1. Determine the EC_{50} value by fitting the data to the following equation (*see* **Note 5** and Fig. 3d):

$$E\,(\%\max) = (E_{\max}\,C^n)\,(\mathrm{EC_{50}}^n + C^n)^{-1} \qquad (1)$$

where C is the concentration of IP_3R ligands, E_{max} is the maximal effect, and n is the Hill coefficient.

2. Calculate $s[IP_3]_i$ (IP_3 concentrations in sample solutions) using the following equation:

$$[\mathrm{IP_3}] = K_d\left\{\Delta R\,(\Delta R_{\max} - \Delta R)^{-1}\right\}^{1/n} \qquad (2)$$

where ΔR is $R - R_{min}$, ΔR_{max} is $R_{max} - R_{min}$, K_d is the apparent dissociation constant obtained in **step 1**, n is the Hill coefficient obtained in **step 1**, and R is the fluorescence emission ratio of samples. R_{min} and R_{max} can be obtained from the emission ratios of U73122-treated and nonstimulated samples with 10 μM IP_3, respectively (*see* below).

3.6 Measurement of IP_3 Concentrations in COS-7 Cells

1. Detach cells from the culture dish using Ca^{2+}-free PBS and transfer them to 1.5 mL microcentrifuge tubes (*see* **Note 6**).
2. Centrifuge the tubes (200 × *g* for 2 min), remove the supernatant, and suspend the cells in 300 μL of HBSS-H (~2 × 10^7 cells/300 μL).

3. Add stimulants, cap tubes, and incubate for appropriate durations. Also prepare nonstimulated cells treated with and without 5 μM U73122 to determine the resting and minimum concentrations of IP_3 (R_{min}).
4. Stop the response by heating in a microwave (500 W, 10 s).
5. Collect the supernatants of these samples by centrifugation (1500 × *g*, 5 min at 4 °C).
6. To determine R_{max}, take an aliquot of the U73122-treated sample (~100 μL) and add IP_3 to a final concentration of 10 μM (*see* **Note 7**).
7. This method allows the IP_3 concentration in the sample solution ($s[IP_3]_i$) to be estimated.

3.7 Estimation of the Dilution Rate of Samples

1. Prepare a cell suspension (~2 × 10^7 cells/300 μL), centrifuge (3000 rpm, 5 min), and remove 200 μL of the supernatant as described in Subheading 3.6.
2. Add 200 μL of 10 μg/mL sulfo-Rhodamine B (SRB-1) to the microcentrifuge tube and mix gently.
3. Centrifuge (3000 rpm, 5 min) the tube and remove the supernatant (SRB-2).
4. Calculate the dilution rate to estimate the cytosolic concentrations of IP_3 using the following equations:

$$V_{cell} = 300 - (F_1/F_2) \times 200 \quad (3)$$

$$R = 300/V_{cell} \quad (4)$$

where V_{cell} is the volume of cells in the suspension, F_1 and F_2 are the fluorescence intensities of SRB-1 and SRB-2, respectively, and R is the dilution rate.

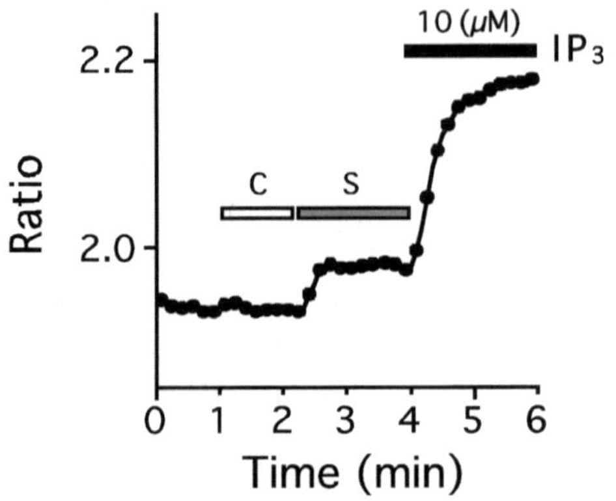

Fig. 4 Measurement of cytosolic IP_3 concentrations using the CFLA-IP_3 assay. Cytosolic samples of nonstimulated (c) and ATP-stimulated (s) cells, and sample c plus 20 μM IP_3 (IP_3) were mixed with the same volume of 60 nM FLL and applied to the IP_3 sensor beads

5. Calculate $c[IP_3]_i$ (cytosolic IP_3 concentration) using the following equation (Fig. 4):

$$c[IP_3]_i = s[IP_3]_i \times R \tag{5}$$

4 Notes

1. Prepared beads can be stored for ~1 week at 4 °C in EQB containing 1% bovine serum albumin.
2. Proteins, such as bovine serum albumin, prevent the attachment of IP_3 sensor beads to glass. Therefore, beads should be washed thoroughly to remove proteins in the solution before they are applied to the chamber.
3. Our system consists of a Nikon TE2000 inverted fluorescence microscope equipped with a Nikon Plan Fluor 40 oil immersion objective (NA 1.3), W-View optics, and a C9100-13 EM-CCD camera (Hamamatsu photonics, Shizuoka, Japan).
4. Figure 3a–c shows changes in the fluorescence ratio with 30 or 100 nM FLL and 0.1, 1, or 10 μM IP_3.
5. Figure 3d shows the concentration–response curve of IP_3 and FRET signal recovery in the presence of 30 and 100 nM FLL calculated using Eq. 1.
6. Prepare tubes to calculate R_{min}, R_{max}, and the estimated dilution rate, in addition to the samples.
7. Use a small amount of a solution containing a high concentration of IP_3, such as 1 μL of a 1 mM solution, to minimize dilution. Alternatively, adjust the dilution rate by adding the same volume of buffer to the other samples.

Acknowledgments

This study was partly supported by the MEXT/Japan Society for the Promotion of Sciences (JSPS) KAKENHI under Grant No. JP16H01433 "Resonance Bio" and Grant No. 26670812 (to A.T.).

References

1. Berridge MJ (1993) Inositol trisphosphate and calcium signalling. Nature 361:315–325
2. Brandish PE, Hill LA, Zheng W, Scolnick EM (2003) Scintillation proximity assay of inositol phosphates in cell extracts: High-throughput measurement of G-protein-coupled receptor activation. Anal Biochem 313:311–318
3. Mathews WR, Gljido DM, Huff RM (1987) Anion-exchange high-performance liquid chromatographic analysis of inositol phosphates. Anal Biochem 168:63–70
4. Singh AK (1992) Quantitative analysis of inositol lipids and inositol phosphates in synaptosomes and microvessels by column chromatography: comparison of the mass analysis and the radiolabelling methods. J Chromatogr 581:1–10
5. Rossi AM, Taylor CW (2011) Analysis of protein-ligand interactions by fluorescence polarization. Nat Protoc *6*:365–387
6. Taouji S, Dahan S, Bossé R, Chevet E (2009) Current screens based on the AlphaScreen™ technology for deciphering cell signalling pathways. Curr Genom *10*:93–101
7. Oura T, Murata K, Morita T, Nezu A, Arisawa M, Shuto S, Tanimura A (2016) Highly sensitive measurement of inositol 1,4,5-trisphosphate by using a new fluorescent ligand and ligand binding domain combination. ChemBioChem 17:1509–1512

Chapter 13

Studying the Effects of Inositol Pyrophosphates in an In Vitro Vesicle–Vesicle Fusion Assay

Mi Sook Sung and Tae-Young Yoon

Abstract

In a recent in vitro vesicle fusion study in which we monitored the fusion of reconstituted SNARE and synaptotagmin-containing proteoliposomes, we discovered that inositol pyrophosphate (5-IP_7) is a potent inhibitor of neuronal exocytosis. We found that the inhibitory effect of 5-IP_7, which is 10 times more potent than those of IP_6 and 1-IP_7, requires direct interaction with synaptotagmin. Here, we provide a detailed protocol for proteoliposome preparation and bulk observation of proteoliposome fusion based on FRET signals.

Key words Single-molecule, SNAREs, Synaoptotagmin-1, Proteoliposome, Membrane fusion, Inositol pyrophosphates

1 Introduction

Exocytosis is a crucial cellular process that releases substances into the extracellular space. In neuronal communication, for instance, neurotransmitters are released from presynaptic termini via exocytosis. In eukaryotes, membrane fusion is an essential process in the last stage of exocytosis that is regulated by a group of proteins including soluble N-ethylmaleimide-sensitive factor attachment protein receptors (SNAREs) and synaptotagmin (Syt) [1]. One group of the SNARE proteins, syntaxin 1 and SNAP-25, are located in the presynaptic plasma membrane where they participate in neurotransmitter release. Another SNARE protein, VAMP2, is located in synaptic vesicle membranes together with the Syt proteins (mainly, Syt1). To facilitate membrane fusion, the three SNARE proteins intertwine with one another to form a four helix-bundle called the SNARE complex [2, 3]. The formation of the SNARE complex not only brings the two membranes into close proximity but it releases up to 65 k_BT of free energy to catalyze membrane fusion [4]. Since SNARE-mediated membrane fusion itself is a Ca^{2+}-independent process, Syt1 serves as a Ca^{2+} switch for

Gregory J. Miller (ed.), *Inositol Phosphates: Methods and Protocols*, Methods in Molecular Biology, vol. 2091, https://doi.org/10.1007/978-1-0716-0167-9_13,

exocytosis, such that the influx of Ca^{2+} ions into presynapses leads to neurotransmitter release on a millisecond time scale. After binding multiple Ca^{2+} ions (four to six), the Syt proteins are thought to accelerate membrane fusion by acting as a bridge between acidic membrane lipids and the SNARE complex.

Inositol phosphates (IPs) are important signaling second messengers found in all eukaryotic organisms from yeast to mammals. Inositol pyrophosphates carry high-energy pyrophosphate bonds and are key regulators of growth and metabolism with a high level of selectivity [5]. In mammals, the most intensively studied inositol pyrophosphate is 5-diphosphoinositol pentakisphosphate (5-PP-[1,2,3,4,6]IP_5; 5-IP_7), which is produced by the phosphorylation of the fully phosphorylated six-carbon inositol (inositol hexakisphosphate; IP_6) at position 5. It should be noted that another IP_7 isomer, 1/3-PP-IP_5 (1-IP_7), also functions in various biological events such as inositol polyphosphate homeostasis as well as cell cycle control [5–7].

Recently, we discovered using an in vitro vesicle–vesicle fusion assay that 5-IP_7 inhibits the vesicle fusion process [8]. In the vesicle–vesicle fusion assay, we reconstituted one of the groups of proteoliposomes with the t-SNARE proteins syntaxin 1 and SNAP-25. We therefore referred to these vesicles as t-SNARE vesicles. We reconstituted the other group of proteoliposomes, which we referred to as v-SNARE vesicles, with VAMP2 and Syt1. By systematically omitting each of these reconstituted proteins, we found that Syt1 is the molecular target through which 5-IP exerts its inhibitory effect on vesicle fusion. We found 5-IP_7 is 10-fold more potent than the other IPs (i.e., IP_6 and 1-IP_7) in inhibiting membrane fusion. While other IPs are recognized as Syt-binding metabolites, our lack of understanding of the relevant molecular mechanisms underlying their activity has aroused suspicion regarding their physiological relevance [9–12]. Our observations reveal 5-IP_7 is a potent, high-affinity inhibitor of synaptic vesicle fusion that directly interacts with Syt1.

In this chapter, we introduce our methods for proteoliposome preparation and for performing our in vitro vesicle fusion assay. Like the one originally developed by the Rothman group, our in vitro vesicle fusion assay is based on the measurement of FRET signals. In our version of the assay, we use DiI and DiD as the FRET donor and acceptor dyes, respectively. This FRET pair has improved fluorescence characteristics, which allow us to image proteoliposomes and their fusion at the level of a single vesicle [13]. Because the details of single-vesicle imaging have been reported several times, we provide here the brief protocols for our single-vesicle fusion assay. The same proteoliposome samples can be used directly for bulk vesicle fusion assay.

2 Materials

2.1 Proteolipids

1. Lipids: 1-palmitoyl-2-oleoyl-sn-glycero-3-phosphocholine (POPC), 1,2-dioleoyl-sn-glycero-3-phospho-L-serine (DOPS), 1,2-dioleoyl-sn-glycero-3-phosphoethanolamine (DOPE), L-α-phosphatidylinositol-4,5-bisphosphate (PIP_2), Cholesterol in chloroform, 1,2-dipalmitoyl-sn-glycero-3-phosphoethanolamine-N-(biotinyl) (Biotinyl Cap PE), 1,1-′-dioctadecyl-3,3,3′,3′-tetramethylindocarbocyanine perchlorate (DiI) in ethanol and 1,1′-dioctadecyl-3,3,3′,3′-tetramethylindocarbocyanine perchlorate (DiD) in ethanol.
2. Lipid mixtures for t-vesicles: (DOPS:POPC:cholesterol:PIP_2: DOPE:DiI = 15:37.5:20:1:25:2.2, molar ratio).
3. Lipid mixtures for v-vesicles: (DOPS:POPC:cholesterol: DOPE:Biotinyl PE:DiD = 3:50.4:20:25:0.1:1.5, molar ratio). Biotin is required for a single-vesicle fusion assay.
4. Fusion buffer (FB): 25 mM HEPES, 100 mM KCl, 5% glycerol, 1 mM Mg^{2+}, 1 mM ATP, pH 7.4 with KOH.
5. 1% OG (1% *N*-octyl-β-D-glucoside) in FB.
6. 1 mM DTT (Dithiothreitol) in FB.
7. Proteins for t-vesicle: (Syntaxin-1A (a.a. 1–288, pGEX-KG), SNAP-25 (a.a. 1–206, pGEX-KG), C-terminal VAMP2 (a.a. 49–96, pGEX-KG))=(1:1:1), (PL ratio 1:500).
8. Proteins for v-vesicle: VAMP2 (a.a. 1–116, pGEX-KG) (PL ratio 1:500), Synaptotagmin-1 (a.a. 51–421, pET-28a) (PL ratio 1:333).
9. For dialysis: SM-2 Biobead, Dialysis tubes, floating racks.

2.2 Vesicle Fusion Assays

1. Each IPs (Inositol pyrophosphates): 1-IP_7, 5-IP_7, and IP_6 [14, 15].
2. Black 96-well plates for spectrometer imaging.
3. Cleaned and PEG-coated quartz slides for total internal reflection fluorescence (TIRF) microscopy imaging.
4. 0.1 mg/ml Neutravidin for surface immobilization of v-vesicles.

3 Methods

3.1 Reconstitution of Proteoliposomes

Proteoliposomes are produced in several ways including dilution, dialysis, and gel chromatography. All of these methods work by diluting a detergent-solubilized protein/lipid mixture and reducing the concentration of the detergent below the critical micelle concentration (CMC). This induces the spontaneous formation of proteoliposomes. We use the mild detergent *n*-octyl-β-D-glucopyranoside (OG), which has a CMC of ~0.7% (w/v). Reconstituted vesicles generated in this way mimic the size of synaptic vesicles, 40–50 nm in diameter, and the plasma membrane composition consisting of PIP_2. We also introduce an imbalance in the distribution of phosphatidylserine (PS) to produce two different types of vesicles: 3 mol% to produce v-vesicles and 15 mol% to produce t-vesicles. This allows us to reproduce the rapid fusion process observed under physiological conditions [13]. The total lipid concentration in the final formulation is 1 mM; the protein-to-lipid (PL) ratios are 1:500 for the SNAREs and 1:333 for Syt1. The formation of vesicles and their resulting size is affected by the initial lipid and detergent concentrations as well as by the PL ratio. Consequently, these conditions also determine the number of copies of each protein per vesicle.

1. Calculate the required amount of lipids based on the molar ratios of lipid composition, (PS:PC:cholesterol:PIP_2:PE:DiI = 15:37.5:20:1:25:2.2) for t-vesicles and (PS:PC:cholesterol:PE:biotin:DiD = 3:50.4:20:25:0.1:1.5) for v-vesicles, keeping the total lipid concentration at 4.5 mM in 0.5 ml of solution.
2. Mix the lipid solutions in glass vials, one for each type of vesicle, and blow them dry with a stream of nitrogen to produce a thin film of lipid.
3. Put the vials under vacuum for 2 h to thoroughly remove any residual organic solvent.
4. Gently vortex the lipid films in 0.5 ml fusion buffer (FB) containing 1% (w/v) OG to solubilize the lipids.
5. Transfer the dissolved lipids from vials to 1.5-ml tubes and agitate them with an Intelli-mixer on rotation mode for 1 h at room temperature.
6. Calculate the required concentration of proteins according to a PL ratio of 1:500 (except for syt1, 1:333).
7. Mix syn-1a, SNAP-25, and C-terminal VAMP2 (cvp) with 1% OG to form precomplexes of t-SNARE proteins stabilized by cvp and agitate the mixture on vibration mode for 1 h at 4 °C (*see* **Note 1**). Hereafter, maintain this mixing mode because of the small volumes involved in each step.
8. While incubating the pre-t-SNARE complexes, mix DiD-labeled lipids and 1 mM DTT separately with Syt1 (1:333) and VAMP2 (1:500) for 1 h (*see* **Note 2**).
9. Add DiI-labeled lipids to the pre-t-SNARE complexes to form t-vesicles and agitate them for 1 h at 4 °C.

10. Transfer the VAMP2/lipid mixture into the Syt1/lipid mixture and agitate it for 1 h at 4 °C.
11. Rapidly dilute the two mixtures (obtained from **steps 9** and **10**) with FB and reduce the concentration of OG to 0.45%, which is below its CMC.
12. Prewet dialysis tubes with FB for 5 min. Prepare two 1-g aliquots of SM-2 Biobead (Bio-Rad) and rinse them twice with distilled water.
13. Add the diluted mixtures to the prewetted tubes and dialyze each in 2 L of FB with the beads overnight at 4 °C. During the dialysis, refresh the buffers with new beads after 2 h (*see* **Note 3**).

3.2 Measurement of Vesicle Fusion Events in Bulk

The ensemble assay that measures rapid changes in FRET during vesicle fusion should be performed for various samples and diverse concentrations of IP_7. We use a fluorescence spectrometer (PerkinElmer) with 96-well plates to record the signals from multiple samples simultaneously for 40 min. We then subtract the background fluorescence intensity from nonreacted vesicles from the intensities of the mixed vesicles. In the bulk assay, the normalized intensity at 5 min is comparable to the docking number in the single-vesicle fusion assay.

1. Calculate the volume of proteoliposomes needed for the final concentration of lipids to reach 50 μM.
2. Set the "Measurement mode" to "By rows," the "Number of plate repeats" to "300" and the "Start plate repeat each" setting to "10 s". This allows the total reaction time to be determined.
3. Prepare t-vesicles with the proper concentrations of calcium ion and IP_7 solutions and then add them serially into the wells (*see* **Note 4**). Apart from the volume of the v-vesicle solution, the remaining volume is filled with fusion buffer.
4. Put the two vesicle solutions, DiI-labeled t-vesicles and DiD-labeled v-vesicles, into each new well to act as background. These two wells should be filled to their final volume with additional fusion buffer.
5. Put the v-vesicles into the wells prepared in **step 3** and measure their fluorescence intensity (554-nm excitation and 664-nm emission) or their FRET (*see* **Note 5**).
6. Determine the vesicle fusion result by subtracting the background intensity measured in **step 4**.

3.3 Measurement of Vesicle Fusion in the Single-Vesicle Assay

Our single-vesicle assay uses total internal reflection (TIR) fluorescence microscopy (customized by IX-71; Olympus). This TIR setup creates an evanescent field that excites fluorescent species within ~100 nm of an imaging surface in a sample chamber. This permits the observation of interactions near the imaging surface using an EM-CCD camera (iXon DU897v; Andor Technology). To immobilize v-vesicles on the imaging surface, we introduce biotinylated lipids in proteoliposomes and tether them to the imaging surface with NeutrAvidin [16]. These surface-bound v-vesicles are labeled with DiD so they can be visualized with a 633-nm laser. We then count the vesicles in a given field-of-view at various concentrations. We keep the sample solutions on ice throughout the experiment to preserve their activity, but carry out the incubation at room temperature until finally adding 80 μl of sample solution to the imaging chamber.

1. Prepare quartz slides cleaned with piranha solution and coated with mPEG and biotin–PEG (Laysan Bio) at a molar ratio of 100:1 (*see* **Note 6**).
2. Assemble a slide with a coverslip to make a microfluidic chamber and seal it with epoxy.
3. Rinse the flow chamber twice with 200 μl of fusion buffer.
4. Inject 0.1 mg/ml NeutrAvidin solution, incubate for 5 min, and wash with fusion buffer.
5. Immobilize v-vesicles on the imaging surface with a 5-min incubation and wash with fusion buffer (*see* **Note 7**).
6. Inject t-vesicles in the presence of IP_7 or Ca^{2+} into a chamber with v-vesicles and wash with fusion buffer after 30 s.
7. Set up the imaging conditions (100-ms exposure) and place the chamber on the TIR microscope stage.
8. Count the number of docked t-vesicles using a 532-nm excitation (*see* **Note 8**).
9. Analyze and average the docking number using custom programs written in IDL (Research Systems) and MATLAB (MathWorks), respectively.

4 Notes

1. C-terminal VAMP2 (a.a. 49–96) inhibits dead-end products by making room in advance for full-length VAMP2 on the opposite vesicles [17]. Even though this in vitro incubation enhances fusion probability, we recommend using coexpressed pre-t-SNARE complexes, referred to as ΔN complexes, to rule out any artificial effects and save reaction time [18].
2. The concentration of OG is maintained at 1% in all solutions containing lipids and proteins until dilution. DTT is used to prevent aggregation of proteins in v-vesicles. Therefore, all buffers for v-vesicles (including dialysis buffer) must always contain 1 mM DTT.

3. Dialysis tubing is semipermeable membranous tubing used for the removal or exchange of small molecules while retaining macromolecules in solution based on differential diffusion. Float 200 μl of sample solution containing 0.45% OG in 2 L of buffer with bio-beads (detergent-adsorbents). Because t- and v-vesicles have different dyes, we prepare dialysis solutions for each, requiring a total of $2 \times 2 \times 2 = 8$ L of buffer.
4. We can characterize the molecular interactions of Syt1 with its two ligands, Ca^{2+}, and 5-IP_7. At a fixed Ca^{2+} concentration, we alter the amount of 5-IP_7 (0–30 μM) or IP_6 (0–1 mM) and check their inhibition of vesicle fusion. Likewise, various Ca^{2+} solutions should be prepared for experiments at a fixed 5-IP_7 concentration. For Ca^{2+}-free conditions, 5 μM EGTA can be used as a Ca^{2+} chelator.
5. To observe different samples at the same time, we recommend a multipipette or automatic machine to speed up and improve the efficiency of the reaction assays. Syt1-mediated vesicle fusion is stimulated within few seconds.
6. Clean quartz slides in piranha solution (200 ml of concentrated sulfuric acid (99% v/v) and 100 ml of hydrogen peroxide (30% v/v)). In detail, heat 200 ml H_2SO_4 solution to 40 °C and immerse the quartz slides in it. Then, add 100 ml of H_2O_2 and bring the temperature up to 120 °C for 10 min. Meanwhile, clean glass coverslips by sonication in 1 M KOH for 30 min. Then, rinse the slides and coverslips in distilled water and sonicate them with MeOH. After that, add an aminosilane solution (MeOH: acetic acid: aminosilane = 100:5:1) for 12 min and rinse with MeOH and distilled water. Dry the slides with a stream of nitrogen and coat them with PEG solution (100 mg/ml mPEG-SVA and biotin-PEG-SVA at a molar ratio of 100:1 in 0.4 M K_2SO_4 and 0.1 M $NaHCO_3$) for 4–6 h at room temperature.
7. The washing step removes floating vesicles. To estimate the number of DiD-labeled vesicles on the surface, count them with varying their concentrations (0–0.3 μM) using 633 nm laser. This allows us to modulate the immobilized v-vesicles per imaging area (45×90 μm^2) to be approximately 30,000 after linear fitting.
8. A DiI–DiD dye pair is used to monitor changes in FRET in real-time experiments. To focus just on the docking of t- and v-vesicles, we recommend immobilizing unlabeled (or DiR-labeled) v-vesicles. This reduces optical cross talk that arises from the direct excitation of DiD into the DiI channel.

Acknowledgments

This work was supported by the National Creative Research Initiative Program (Center for Single-Molecule Systems Biology to T.-Y.Y).

References

1. Rothman JE (1994) Mechanisms of intracellular protein transport. Nature 372:55–63. https://doi.org/10.1038/372055a0
2. Sudhof TC (2004) The synaptic vesicle cycle. Annu Rev Neurosci Palo Alto 27:509–547
3. Jahn R, Scheller RH (2006) SNAREs – engines for membrane fusion. Nat Rev Mol Cell Biol 7: nrm2002. https://doi.org/10.1038/nrm2002
4. Gao Y, Zorman S, Gundersen G et al (2012) Single reconstituted neuronal SNARE complexes zipper in three distinct stages. Science 337:1340–1343. https://doi.org/10.1126/science.1224492
5. Hatch AJ, York JD (2010) SnapShot: inositol phosphates. Cell 143:1030–1030.e1. https://doi.org/10.1016/j.cell.2010.11.045
6. Mulugu S, Bai W, Fridy PC et al (2007) A conserved family of enzymes that phosphorylate inositol hexakisphosphate. Science 316:106–109. https://doi.org/10.1126/science.1139099
7. Lee Y-S, Mulugu S, York JD, O'Shea EK (2007) Regulation of a Cyclin-CDK-CDK inhibitor complex by inositol pyrophosphates. Science 316:109–112. https://doi.org/10.1126/science.1139080
8. Lee T-S, Lee J-Y, Kyung JW et al (2016) Inositol pyrophosphates inhibit synaptotagmin-dependent exocytosis. Proc Natl Acad Sci U S A 113:8314–8319. https://doi.org/10.1073/pnas.1521600113
9. Fukuda M, Aruga J, Niinobe M et al (1994) Inositol-1,3,4,5-tetrakisphosphate binding to C2B domain of IP4BP/synaptotagmin II. J Biol Chem 269:29206–29211
10. Llinás R, Sugimori M, Lang EJ et al (1994) The inositol high-polyphosphate series blocks synaptic transmission by preventing vesicular fusion: a squid giant synapse study. Proc Natl Acad Sci U S A 91:12990–12993
11. Fukuda M, Moreira JE, Lewis FM et al (1995) Role of the C2B domain of synaptotagmin in vesicular release and recycling as determined by specific antibody injection into the squid giant synapse preterminal. Proc Natl Acad Sci U S A 92:10708–10712
12. Sasakawa N, Ohara-Imaizumi M, Fukuda M et al (2011) Dissociation of inositol polyphosphates from the C2B domain of synaptotagmin facilitates spontaneous release of catecholamines in adrenal chromaffin cells. A suggestive evidence of a fusion clamp by synaptotagmin. Neuropharmacology 60:1364–1370. https://doi.org/10.1016/j.neuropharm.2011.03.005
13. Lee H-K, Yang Y, Su Z et al (2010) Dynamic Ca^{2+}-dependent stimulation of vesicle fusion by membrane-anchored synaptotagmin 1. Science 328:760–763. https://doi.org/10.1126/science.1187722
14. Capolicchio S, Thakor DT, Linden A, Jessen HJ (2013) Synthesis of unsymmetric diphospho-inositol polyphosphates. Angew Chem Int Ed 52:6912–6916. https://doi.org/10.1002/anie.201301092
15. Capolicchio S, Wang H, Thakor DT et al (2014) Synthesis of densely phosphorylated Bis-1,5-diphospho-myo-inositol tetrakisphosphate and its enantiomer by bidirectional P-anhydride formation. Angew Chem Int Ed 53:9508–9511. https://doi.org/10.1002/anie.201404398
16. Roy R, Hohng S, Ha T (2008) A practical guide to single-molecule FRET. Nat Methods 5:507–516. https://doi.org/10.1038/nmeth.1208
17. Pobbati AV, Stein A, Fasshauer D (2006) N-to C-terminal SNARE complex assembly promotes rapid membrane fusion. Science 313:673–676
18. Stein A, Radhakrishnan A, Riedel D et al (2007) Synaptotagmin activates membrane fusion through a Ca^{2+}-dependent trans interaction with phospholipids. Nat Struct Mol Biol 14:904–911

Part III

Chemical Tools

Chapter 14

Engineering Protease-Resistant and Highly Active Phytases

Canfang Niu, Peilong Yang, and Bin Yao

Abstract

Phytases can catalyze the hydrolysis of indigestible phytate and releases the usable phosphorus. Protease resistance and high activity of enzymes facilitate their biotechnological and medical application. Here we described a genetic manipulation method to improve enzyme tolerance to pepsin, trypsin, and low pH by optimizing the residual side chain of trypsin- and pepsin-sensitive HAP phytase YeAPPA from *Yersinia enterocolitica*.

Key words HAP phytase, Pepsin resistance, Trypsin resistance, Low PH, Optimizing the residual side chain, High activity

1 Introduction

The inositol phosphates were obtained through hydrolyzing phytate by active phytases [1]. However, the susceptibility to proteases usually reduces the in vivo activity and stability of phytases and decreases nutrient utilization efficiency. The thermostability, pH tolerance, and catalytic activity of phytases have previously been achieved by glycosylation, the reduced structure flexibility, the optimization of protein surface properties, and the presence of disulfide bond, hydrogen bonds and ionic interactions [2–7]. Thermostability and pepsin resistance under acidic condition are not necessarily coupled [8, 9]. We are able to increase the thermal and pepsin resistance of the F89S mutant of YeAPPA from *Y. enterocolitica* and the thermostability and resistance to pepsin and trypsin of the K226H and F89S/K226H mutants of YeAPPA (Figs. 1 and 2).

Gregory J. Miller (ed.), *Inositol Phosphates: Methods and Protocols*, Methods in Molecular Biology, vol. 2091, https://doi.org/10.1007/978-1-0716-0167-9_14, © Springer Science+Business Media, LLC, part of Springer Nature 2020

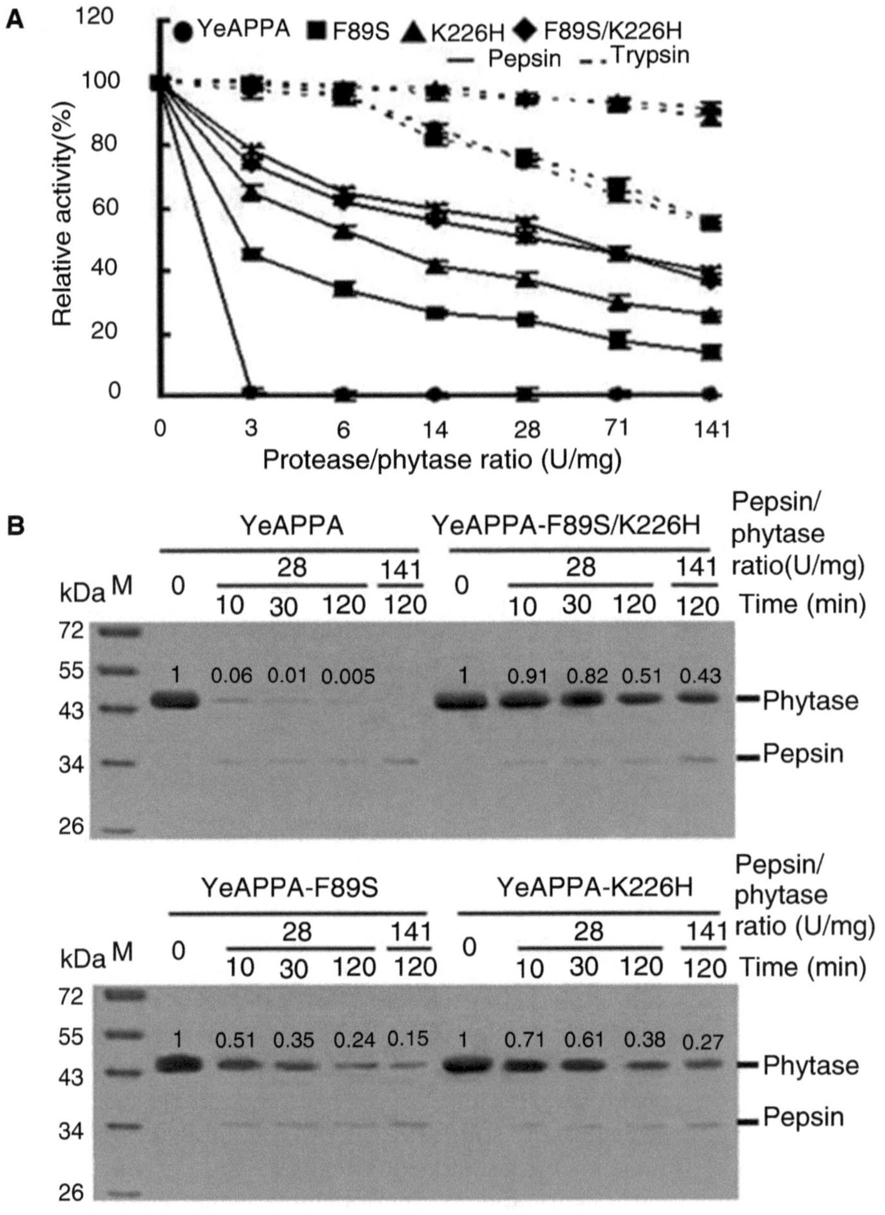

Fig. 1 Proteolytic resistance of YeAPPA and its mutant phytases. A Evaluation of resistance to pepsin pepsin at pH 2 and trypsin at pH 7 for 2 h at 37 °C. B SDS-PAGE analysis of the proteolytic products of YeAPPA and its mutants degraded by pepsin at different ratios. The phytase band intensity was assessed by using ImageJ software. (Reprinted (adopted) with permission from Niu C, Yang P, Luo H, Huang H, Wang Y, et al. (2017) Engineering of *Yersinia* phytases to improve pepsin and trypsin resistance and thermostability and application potential in the food and feed industry. J Agric Food Chem 65:7337–7344. Copyright (2017) America Chemical Society)

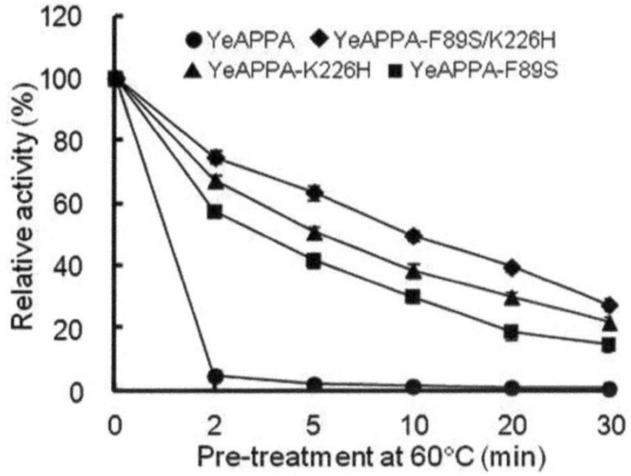

Fig. 2 Thermostability of YeAPPA and its mutants at 60 °C. (Reprinted (adopted) with permission from Niu C, Yang P, Luo H, Huang H, Wang Y, et al (2017) Engineering of *Yersinia* phytases to improve pepsin and trypsin resistance and thermostability and application potential in the food and feed industry. J Agric Food Chem 65:7337–7344. Copyright (2017) America Chemical Society)

2 Materials

Store all strains, vectors, DNA fragments, restriction endonucleases, T4 ligase, and DNA polymerase at −20 °C. Store substrate phytate, pepsin, and trypsin at 4 °C. The chemicals are of analytical grade.

2.1 Phytase Mutagenesis

1. The cDNA fragment of YeAPPA: GenBank accession no. GU936684.
2. Pfu DNA polymerase.
3. Oligonucleotide primers specific for YeAPPA and its mutants (*see* **Note 1**).
4. DNA purification kit.
5. Restriction endonucleases *Eco*RI and *Not*I.
6. pEASY-T3 vector.
7. T4 DNA ligase.
8. *Escherichia coli* Trans1-T1 cells.

2.2 Phytase Production

1. pPIC9γ vector.
2. Restriction endonucleases *Eco*RI, *Not*I, and *Bgl*II.
3. T4 DNA ligase.
4. *Pichia pastoris* GS115 competent cells (*see* **Note 2**).

5. 1 M sorbitol: Dissolve 182 × *g* sorbitol to a final volume of 1 L deionized water, sterilized by filtration.
6. MD plate: Dissolve 20 g agarose in a final volume of 800 mL distilled water for 15 min at 121 °C, and then add 100 mL sterile-filtered 10× YNB to a final concentration of 13.4 g/L, 100 mL sterile-filtered 10× glucose to a final concentration of 20 g/L and 0.2 mL sterile-filtered 500× biotin to a final concentration of 0.0004 g/L in the sterile bench after the temperature drops to 60 °C.
7. BMGY media: Dissolve 10 g yeast extract, 20 g peptone and 10 mL glycerol in a final volume of 900 mL distilled water, sterilize it at 121 °C for 15 min, and then supplement 100 mL sterile-filtered 10× YNB to a final concentration of 13.4 g/L and 0.2 mL sterile-filtered 500× biotin to a final concentration of 0.0004 g/L in the sterile bench after the temperature drops to 60 °C.
8. BMMY media: Dissolve 10 g yeast extract and 20 g peptone in a final volume of 895 mL distilled water, sterilize it at 121 °C for 15 min, and then supplement 100 mL sterile-filtered 10× YNB to a final concentration of 13.4 g/L, 0.2 mL sterile-filtered 500× biotin to a final concentration of 0.0004 g/L and 5 mL methanol in the sterile bench after the temperature drops to 60 °C.
9. Vivaflow 200 membrane with a 10-kDa cutoff.
10. Protein liquid chromatography system consisting of nickel-nitrilotriacetic acid and diethylaminoethyl columns are purchased from Qiagen, Hilden and GE Healthcare, Munich, Germany, respectively.
11. Protein assay kit II including dye reagent concentration and bovine serine albumin standard are purchased from Bio-Rad, Hercules, CA.

2.3 Pepsin and Trypsin Resistance Assay

1. Purified YeAPPA, F89S, K226H, and F89S/K226H phytases.
2. 2827 U/mg pepsin: Dissolve 1 g pepsin (1:3000) in a final volume of 1061 mL 0.25 M glycine–HCl (pH 2) (*see* **Note 3**).
3. 2839 U/mg trypsin: Dissolve 1 g trypsin (1:250) in a final volume of 88 mL 0.25 M Tris–HCl (pH 7) (*see* **Note 4**).
4. 1.5 mM sodium phytate: Dissolve 1386 g sodium phytate in a final volume of 1 L 0.25 M Tris–HCl (pH 5).
5. 10% trichloroacetic acid: Dissolve 100 g trichloroacetic acid in a final volume of 1 L deionized water.
6. Color reagent: 1% ammonium molybdate, 3.2% sulfuric acid, and 7.2% ferrous sulfate (*see* **Note 5**).
7. Spectrophotometer.

8. 12% SDS-PAGE precast gel, 25× SDS-PAGE precast-gel running buffer, and 4× SDS lysis buffer with β-mercaptoethanol: Purchase from Beijing Solarbio Science and Technology Co. Ltd, China.
9. Coomassie Brilliant Blue (CBR-250).

3 Methods

3.1 Phytase Mutagenesis

1. Predict the solvent surface accessibility of theoretical pepsin/trypsin cleavage sites in YeAPPA phytase by using the I-TASSER program (http://zhanglab.ccmb.med.umich.edu/I-TASSER/).
2. Substitute the surface exposed residues, pepsin cleavage site F89 and pepsin/trypsin site K226, with S and H, respectively, by overlap-extension PCR (*see* Ref. [10]) using the *YeAPPA* phytase gene as the DNA template with specific primers (*see* **Note 1**).
3. Insert the variant genes into the pEASY-T3 vector.
4. Sequence the introduced DNA fragments in the recombinant pEASY-T3 vector.

3.2 Phytase Production

1. Digest YeAPPA and its mutants harboring the correct mutations with restriction endonucleases *Eco*RI and *Not*I at 37 °C for 1 h.
2. Individually insert the wild-type and mutant phytase gene fragments into the pPIC9γ expression vector by T4 DNA ligase at 16 °C for 10 min.
3. Linearize 100 μg of pPIC9γ recombination vector with 5 μL *Bgl*II at 37 °C for 1 h.
4. Transform 1–20 μg of linearized pPIC9γ recombination vector to 80 μL of *P. pastoris* GS115 competent cells by using 0.2 cm electric conversion cup at voltage of 2000 V for about 5 ms.
5. Mix the transformation solution with 400 μL of 1 M precooled sorbitol.
6. Spread the mixture on the MD plates.
7. Incubate the MD plates at 30 °C for 2–4 days.
8. Pick a single colony, grow it in 5 mL BMGY media at 30 °C for 2 days, harvest the cells by centrifugation at 3000 rpm for 5 min, and grow them in 5 mL BMMY media containing 25 μL methanol as inducer at 30 °C for 2 days.
9. Inoculate the highest-expressing colony in 200 mL BMGY media, grow it in 200 mL BMMY containing 1 mL methanol as inducer at 30 °C for 2 days, harvest the cultured cells by

centrifugation at 3000 rpm for 5 min, and grow them in 200 mL BMMY media at 30 °C for 5 days.

10. Centrifuge the culture solution at 5000 rpm for 5 min and collect the culture supernatant.
11. Concentrate the culture supernatant by a Vivaflow 200 membrane with a 10-kDa cutoff.
12. Load the crude enzymes onto a fast protein liquid chromatography system consisting of nickel-nitrilotriacetic acid and diethylaminoethyl columns.
13. Elute the proteins by a linear gradient of 0–1 M NaCl at a flow rate of 2 mL/min and pool the fractions with phytase activity.
14. Determine total protein concentration by a Bio-Rad protein assay kit.

3.3 Pepsin and Trypsin Resistance Assay

1. Dilute the wild-type and variant phytases in 0.25 M Tris–HCl (pH 5) to a concentration of 3.2 U/mg.
2. Incubate YeAPPA and its variants (3.2 U/mg) with either pepsin (2827 U/mg) in 0.25 M glycine–HCl (pH 2) or trypsin (2839 U/mg) in 0.25 M Tris–HCl (pH 7) at protease/phytase ratios of 3–141 U/mg at 37 °C for 2 h.
3. Aliquote the enzyme solutions into two part for the residual phytase activity assay and the proteolysis product analysis.
4. Residual phytase activity assay: Mix 50 μL of pepsin and trypsin digested phytases with 950 μL of 1.5 mM sodium phytate as the substrate, incubate it at 37 °C for 30 min, then add 1 mL of 10% trichloroacetic acid to stop the reaction and develop it with 2 mL of color reagent (*see* Ref. [11]). Measure the absorbance at 700 nm and calculate the enzyme activity (*see* **Note 6** and 7).
5. Proteolysis product analysis: Mix the pepsin proteolysis mixtures with SDS lysis buffer, boil them for 10 min, load the protein lysis solution on the SDS-PAGE gel with protein standard, electrophorese at 25 mA till the dye front reaches the bottom of the gel. Following electrophorese, stain the gel with Coomassie Brilliant Blue (CBR-250). Determine the phytase band intensity by using ImageJ software.

3.4 Thermostability Assay

1. Mix the YeAPPA and its mutant phytases with 1.5 mM sodium phytate and incubate them at 37 °C for 30 min in 0.25 M buffer systems: pH 1–3.5 glycine–HCl, pH 3.5–6 sodium acetate–acetic acid, pH 6–8.5 Tris–HCl, and pH 8.5–12 glycine–NaOH.
2. Measure the activities of each phytase treated with various pHs as in Subheading 3.3, **step 4** (*see* **Note 6**) and determine their pH optima.

3. Preincubate the YeAPPA and its mutant phytases at the optimal pH and 60 °C in the absence of substrate for the desired times (0, 2, 5, 10, 20, and 30 min).
4. Quickly put the phytases on ice after heat treatment.
5. Measure the residual activities of the heat-treated phytases as in Subheading 3.3, **step 4** (*see* **Notes 6** and 7).

4 Notes

1. Oligonucleotide primers specific for YeAPPA and its mutants include YeAPPA-Forward:5′-CGC*GAATTC*GCACCGCTTGCAGCACAATCTAC-3′, YeAPPA-Reverse:5′-GAT*GCGGCCGC*TTAAATATGGCAGGCTGGCTCG-3′; YeAPPA-F89S-Forward:5′-TGATGGGGGGAAGCTATGGTGATTATTTC-3′, YeAPPA-F89S-Reverse:5′-GAAATAATCACCATAGCTTCCCCCCATCA-3′; YeAPPA-K226H-Forward:5′-CAAATGAAATTCATGTAAACGAAGAAGGTA-3′, YeAPPA-K226H-Reverse:5′-TACCTTCTTCGTTTACATGAATTTCATTTG-3′.
2. *Pichia pastoris* GS115 competent cells can be prepared as follows. Inoculate a single colony of yeast in 10 mL of YPD liquid medium at 30 °C overnight. Transfer 300 μL of the overnight culture to a flask containing 200 mL of BMGY liquid medium and grow it at 30 °C to OD_{600} = 0.9–1.3. Centrifuge the culture solution at 5000 rpm for 5 min at 4°°C and discard the supernatant. Rinse the cells with 200 ml of precooled sterile ultrapure water twice and then rinse them with 40 mL of 1 M precooled sorbitol. Resuspend the rinsed cells in 400 μL of 1 M precooled sorbitol.
3. One unit (U) of pepsin was defined as a 0.001 change in absorbance at 280 nm/min at 37 °C and pH 2.0 with hemoglobin as the substrate.
4. One unit (U) of trypsin was defined as a 0.001 change in absorbance at 253 nm/min at 37 °C and pH 7.0 with Na-Benzoyl-L-arginine ethyl ester (BAEE) as the substrate.
5. Firstly prepare ammonium molybdate–sulfuric acid solution: dissolve 10 g of ammonium molybdate in 100 mL of deionized water, supplement it with 32 mL of sulfuric acid and fill it up to a final volume of 1 L with deionized water. Dissolve 7.2 g of ferrous sulfate in 100 mL of ammonium molybdate–sulfuric acid solution to make color reagent.

6. Define one unit (U) of activity as the amount of enzyme that liberates 1 μmol of phosphate per minute under assay conditions.
7. Regard the phytase activity toward 1.5 mM sodium phytate at 37 °C for 30 min as 100%.

References

1. Jiang H, Ma J, Wu M (2004) Study on producing inositol triphosphate through the phytase extraction of wheat bran. Food Sci 25:124–127
2. Liao Y, Li C, Chen H, Wu Q, Shan Z, Han X (2013) Site-directed mutagenesis improves the thermostability and catalytic efficiency of *Aspergillus niger* N25 phytase mutated by I44E and T252R. Appl Biochem Biotechnol 171:900–915
3. Wu T, Chen C, Cheng Y, Ko T, Lin C, Lai H et al (2014) Improving specific activity and thermostability of *Escherichia coli* phytase by structure-based rational design. J Biotechnol 175:1–6
4. Zhang W, Lei X (2008) Cumulative improvements of thermostability and pH-activity profile of *Aspergillus niger* PhyA phytase by site-directed mutagenesis. Appl Microbiol Biotechnol 77:1033–1040
5. Oda K (2012) New families of carboxyl peptidases: serine-carboxyl peptidases and glutamic peptidases. J Biochem 151:13–25
6. Ahn J, Cao M, Yu Y, Engen JR (2013) Accessing the reproducibility and specificity of pepsin and other aspartic proteases. Biochim Biophys Acta 1834:1222–1229
7. Niu C, Luo H, Shi P, Huang H, Wang Y, Yang P et al (2015) *N*-Glycosylation improves the pepsin resistance of histidine acid phosphatase phytases by enhancing their stability at acidic pHs and reducing pepsin's accessibility to its cleavage sites. Appl Environ Microbiol 82:1004–1014
8. Kumar K, Dixit M, Khire J, Pal S (2013) Atomistic details of effect of disulfide bond reduction on active site of Phytase B from *Aspergillus niger*: A MD Study. Bioinformation 9:963–967
9. Zhao Q, Liu H, Zhang Y, Zhang Y (2010) Engineering of protease-resistant phytase from *Penicillium* sp.: high thermal stability, low optimal temperature and pH. J Biosci Bioeng 110:638–645
10. Niu C, Yang P, Luo H, Huang H, Wang Y et al (2017) Engineering the residual side chains of HAP phytases to improve their pepsin resistance and catalytic efficiency. Sci Rep 7:46871
11. Niu C, Yang P, Luo H, Huang H, Wang Y et al (2017) Engineering of *Yersinia* phytases to improve pepsin and trypsin resistance and thermostability and application potential in the food and feed industry. J Agric Food Chem 65:7337–7344

Chapter 15

Direct Cloning Method for Expression of Recombinant Proteins with an Inositol Hexakisphosphate Inducible Self-Cleaving Tag

Keehwan Kwon, Marco Biancucci, Patrick J. Woida, and Karla J. F. Satchell

Abstract

Protein purification is the most basic and critical step for protein biophysical and biochemical studies to understand its function and structure. Various fusion tags and proteases have been developed and assembled in expression and purification system. However, it is one of the fields that continues to innovate to develop improved systems that are more efficient, simpler, and less expensive. An efficient self-cleavage C-terminal fusion system was developed using the inositol hexakisphosphate-inducible *Vibrio cholerae* $MARTX_{Vc}$ toxin cysteine protease domain (CPD). CPD fusion proteins are expressed from the T7 promoter and purified using a 6xHis-tag with immobilized-metal affinity chromatography. The C-terminal CPD-tag is removed by self-cleavage at the final purification stage. Here, we describe an efficient cloning method using Gibson assembly, followed by expression and purification of tagless recombinant proteins of interest using CPD self-cleavage.

Key words Cloning, Gibson assembly, Protein expression, Protein purification, SDS-PAGE, CPD, Self-cleavage fusion tag, Inositol hexakisphosphate, IMAC, IP6, $InsP_6$

1 Introduction

Multifunctional-autoprocessing Repeats-in-Toxins (MARTX) toxins are large bacterially secreted proteins of >3500 amino acids [1]. The best studied MARTX toxin for function is the $MARTX_{Vc}$ toxin, encoded by *Vibrio cholerae* gene *rtxA* (*vc1451*). All MARTX toxins are thought to self-assemble on the surface of eukaryotic cells to form a pore and then translocate the central core region across the plasma membrane [2]. One domain transferred is the autoprocessing cysteine protease domain (CPD). Within the eukaryotic cell cytosol, the CPD has high binding affinity for inositol hexakisphosphate ($InsP_6$), a molecule present at high concentration in eukaryotic cell cytosol [3, 4]. $InsP_6$ binds to a basic,

Gregory J. Miller (ed.), *Inositol Phosphates: Methods and Protocols*, Methods in Molecular Biology, vol. 2091, https://doi.org/10.1007/978-1-0716-0167-9_15,

lysine-rich pocket that is separated on the opposite side from the catalytic His-Cys dyad and substrate binding S1 site by a structure labeled the β-flap [3, 5]. Upon binding to $InsP_6$, the CPD completes folding to enter a thermodynamically stable state that locks the substrate peptide into the S1 site and orientates the scissile bond toward the catalytic cysteine [5, 6]. This activates the cysteine nucleophile leading to the cleavage of a Leu-Xaa scissile bond [5, 7].

Following cleavage, the CPD recycles the $InsP_6$ molecule and then converts to a distinct conformation state that is more thermodynamically stable, but has a low binding affinity for new $InsP_6$ molecules [5]. This suggests a functional mechanism in which, following translocation, CPD binds $InsP_6$ and cleaves a *cis* site at its own N-terminus. The enzyme can be reactivated to cleave in trans by cooperative binding of a new peptide substrate and $InsP_6$ molecule [6].

This unique autoprocessing mechanism has been exploited for recombinant protein production in *Escherichia coli*, independently developed by two different groups [8, 9]. Since bacteria do not generally utilize inositol signaling, the CPD purifies from *E. coli* as an intact, uncleaved protein [10]. Thus, a recombinant protein consisting of an N-terminal protein fused to a C-terminal CPD-six histidine tag (called CPD-6xHis or simply the "CPD-tag") is easily purified at full-length using immobilized metal affinity chromatography (IMAC) [8, 11]. Subsequently, the CPD-tag is removed by autoprocessing induced by incubation in $InsP_6$, with recovery of the protein of interest (POI) by passage again over the metal affinity column or by directly cleaving on the column using an $InsP_6$-containing buffer. A noted feature of this system is that the addition of the CPD-tag can in some cases function as a C-terminal solubility tag, improving recovery of soluble protein during overexpression [11].

This chapter specifically describes two recently developed pCPD vectors for easy insertion of gene sequences to generate CPD gene fusions. The pCPD vectors were constructed in the backbone of plasmid pMCSG58, which includes rare codons that enhance successful expression [12](*see* Fig. 1). Also carried over from pMCSG58, the pCPD vectors are engineered to be compatible with classical blunt-end restriction enzyme cloning and ligation independent cloning (LIC) [11]. Here, we introduce a detailed Gibson assembly method for fast and efficient cloning of PCR amplified DNA using the pCPD vector modified to carry the *ccdB* toxin gene. This method is advantageous as no further purification or treatment of the PCR product or digested vector backbone are required prior to Gibson assembly, providing a major advantage over classical and ligation-independent cloning strategies. This method will detail PCR amplicons, preparation of vector backbone, and Gibson assembly setup. Elements of gene expression, protein purification, and tag removal unique to CPD-tag system are detailed.

2 Materials

2.1 Preparation of Linearized pCPD Vector Backbone

1. Purified supercoiled pCPD-*ccdB* vector DNA (*see* **Note 1** and Fig. 1).
2. SmaI restriction endonuclease and 10× NEB CutSmartBuffer (New England BioLabs, Ipswich, MA, USA).
3. Supplies and reagents necessary for agarose gel electrophoresis including agarose, buffers, and stain [13, 14].
4. 10× Agarose Gel Loading Buffer.
5. 1 kb DNA Ladder.

2.2 Polymerase Chain Reaction (PCR)

1. Purified template DNA containing gene of interest (genomic DNA, cDNA, or plasmid).
2. Custom forward and reverse PCR primers (Integrated DNA Technologies, Coralville, IA, USA). Separately prepare dilution of forward and reverse primers at a concentration of 5 μM and store at −20 °C (*see* **Notes 2–5**).
3. 100 mM dNTP set (dATP, dCTP, dGTP, dTTP).
4. Phusion® High-Fidelity DNA Polymerase (New England Biolabs).
5. MinElute PCR Purification Kit (Qiagen, Valencia, CA).
6. 5× isothermal reaction buffer: 500 mM Tris–HCl pH 7.5, 50 mM $MgCl_2$, 1 mM dNTPs, 50 mM dithiothreitol (DTT),

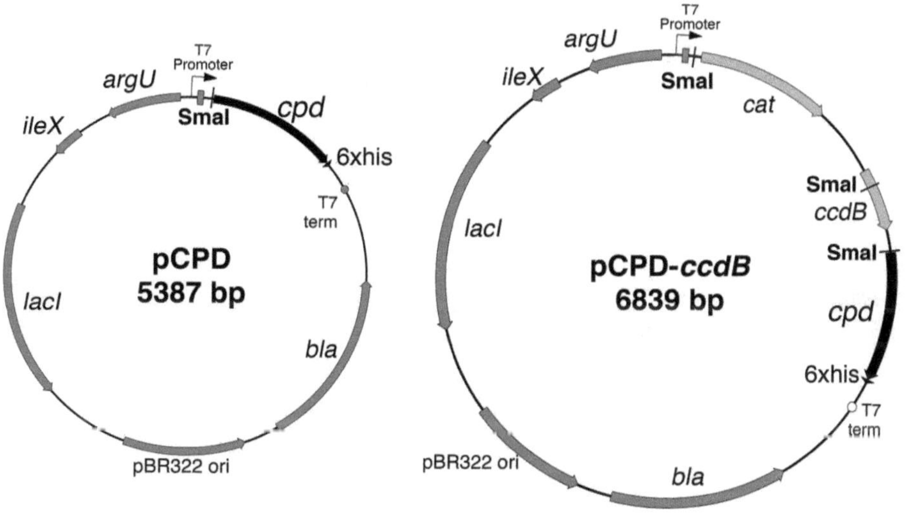

Fig. 1 Map of pCPD vectors. Gene designations are *cpd-6xhis*, *V. cholerae* cysteine protease domain. *bla*, ampicillin resistance; *cat*, chloramphenical resistance; *ccdB*, toxin gene; rare codon tRNAs, *ileX* and *argU*; *lacI*, lactose inducer. Both vectors are linearized with SmaI for gene insertion. pCPD-*ccdB* has three SmaI sites that removes the *cat-ccdB* cassette and cleaves the insert

Table 1
5× Isothermal reaction buffer

Components	Volume
1 M Tris–HCl pH 7.5	3 ml
2 M $MgCl_2$	150 μl
100 mM dGTP	60 μl
100 mM dATP	60 μl
100 mM dTTP	60 μl
100 mM dCTP	60 μl
1 M DTT	300 μl
PEG-8000	1.5 g
100 mM NAD^+	300 μl
Autoclaved water	Bring up the volume to 6 ml

25% (w/v) polyethylene glycol (PEG-8000), 5 mM β-nicotinamide adenine dinucleotide (NAD^+). Assemble from premade sterile stock solutions according to Table 1. Store at −20 °C in 320 μl aliquots (*see* **Note 6**).

7. Assembly master mixture: 1.33× isothermal reaction buffer, 0.0053 U T5 exonuclease, 0.033 U Phusion® High-Fidelity DNA polymerase, 5.33 U Taq DNA ligase. All enzymes are from New England Biolabs. Assemble according to Table 2. Stored at 20 °C in 15 μl aliquots (*see* **Note 6**).
8. *E. coli* DH5α competent cells or other *gyrA*$^+$ *E. coli* used for DNA cloning (*see* **Note 7**).
9. T7 promoter and T7 terminator sequencing primers (*see* **Note 8**).

2.3 CPD Fusion Overexpression of Protein in Bacteria

1. *E. coli* BL21 (DE3) competent cells (*see* **Note 7**).
2. Bacterial broth culture media prepared from broth base (*see* **Note 9**).
3. Ampicillin antibiotic stock solution: Dissolve 1.0 g ampicillin sodium salt in ddH_2O for final volume of 10 ml. Sterilize by passage over a 0.22 μ syringe filter and store at 4 °C. Add ampicillin to broth media at 1 ml per 1-L media for final concentration of 100 μg per ml.
4. Isopropyl-β-D-thiogalactoside (IPTG): Prepare 100 mM solution by dissolving 0.24 g IPTG powder in distilled water for final volume of 10 ml. Sterilize by passage over a 0.22 μ syringe filter and store at 4 °C.

Table 2
Assembly master mixture

Components	Volume
5× isothermal reaction buffer	320 μl
NEB T5 exonuclease (10 U/μl)	0.64 μl
NEB Phusion® High-Fidelity DNA polymerase	20 μl
Taq DNA ligase (40 U/μl)	160 μl
Autoclaved water	699.4 μl

2.4 CPD-Fusion Protein Purification and Autoprocessing

1. IMAC Protein purification reagents (*see* Subheading 3.5).
2. 100 mM phytic acid ($InsP_6$) solution. Dissolve 0.66 g phytic acid sodium salt hydrate from rice (Sigma-Aldrich) in final volume of 10 ml ddH_2O.
3. Buffer A1 for CPD autoprocessing: 50 mM Tris–HCl pH 8.0, 500 mM NaCl prepared by dilute 25 ml 2 M Tris–HCl pH 8.0 and 100 ml 5 M NaCl stock solutions to 1-L in a graduated cylinder (*see* **Note 10**).
4. Imidazole containing buffers A2, A3, and A4. Prepare as for Buffer A1 except adding 10 ml 2 M imidazole for 20 mM imidazole Buffer A2, 25 ml for 50 mM imidazole Buffer A3, and 250 ml for preparation of 500 mM imidazole Buffer A4 (*see* **Note 11**).
5. 2× Laemmli Sample buffer: 120 mM Tris–HCl pH 6.8, 4% SDS, 10% glycerol, 0.02% bromophenol blue.
6. Reagents and supplies for sodium dodecyl sulfate-polyacrylamide electrophoresis (SDS-PAGE) including gels, buffers, and Coomassie blue stain and destain following established protocols [15, 16].

2.5 Equipment

1. Thermal cycler.
2. Small size agarose gel electrophoresis system with included casting tray.
3. UV transilluminator.
4. NanoDrop spectrophotometer.
5. Fast-performance liquid chromatography (FPLC) system.
6. SDS-PAGE mini-gel system.

3 Methods

3.1 Preparation of Linearized pCPD Vector Backbone

1. The overall process of Gibson assembly cloning is depicted in Fig. 2. The pCPD vectors used to generate fusions of the gene of interest to *cpd-6xhis* are linearized using SmaI restriction enzyme digestion. For the pCPD-*ccdB* vector (shown in Fig. 1), the *ccdB* toxin gene is located between two SmaI cleavage such that the gel purification step is not necessary since undigested or partially digested vector will not yield viable bacterial colonies (*see* **Notes 1** and **12**).
2. Transfer 5 μg pCPD-*ccdB* vector to a PCR tube and bring the volume to 41 μl with autoclaved ddH_2O water. Add 5 μl 10× NEB CutSmart Buffer and mix well.
3. Add 4 μl SmaI (20 U/μl) and incubate the reaction mixture at 25 °C for 4 h (*see* **Note 13**). Then, incubate at 65 °C for 20 min to inactivate SmaI. The cleavage reaction can be carried out in a programmed thermal cycler followed by holding incubation at 4 °C until used.
4. Linearization of the plasmid vector backbone is confirmed by running a sample of the reaction using agarose gel electrophoresis. Fill the agarose gel apparatus with 1× TAE running buffer containing 0.5 μg/ml ethidium bromide and place 0.9% agarose gel in apparatus. Mix 9 μl of SmaI digestion reaction with 1 μl 10× Blue-Juice, load into well of the gel, and separate the linearized plasmid from uncut vector at 80 V for 60 min or until the blue dye front is 2/3 of distance across the gel. Visualize DNA bands using a UV transilluminator and estimate size by comparison to a 1 kb ladder run in parallel. The digested pCPD-*ccdB* vector will show two bands, the linearized pCPD vector of 5.4 kb and the excised *ccdB* fragment of 1.8 kb. Photograph and discard gel. If digestion is incomplete, add another 4 μl SmaI to remaining 41 μl enzyme reaction, incubate again, and retest for cleavage (*see* **Note 13**).
5. If digestion is verified complete by analysis of fragments on an agarose gel, no further gel purification of the remainder of the linearized vector reaction is necessary. The concentration can be based on the molar concentration of plasmid added to the reaction (*see* **Note 14**).
6. The reaction can be stored at −20 °C for a few months.

3.2 Preparation of DNA Inserts by Polymerase Chain Reaction (PCR)

1. Design and purchase primers for amplification of gene sequences with extensions for insertion into linearized pCPD vector (*see* Fig. 2 and **Note 3**).
2. Set up program on the thermal cycler for PCR as described in Table 3 (*see* **Note 15**).

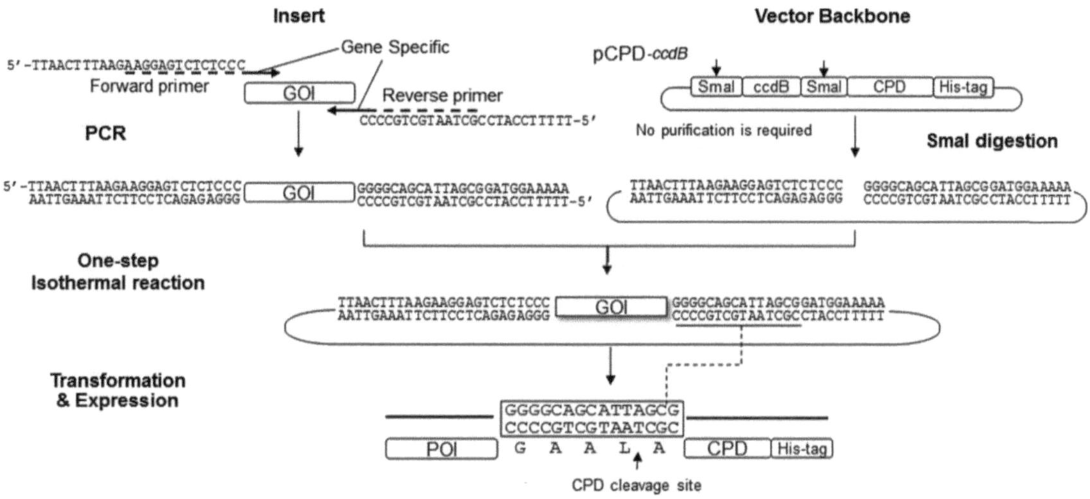

Fig. 2 Overview of Gibson assembly reaction with pCPD-*ccdB* and PCR product. A pair of PCR primers are designed based on gene of interest (GOI) sequences. GOI specific regions of each primer are designed to introduce an ATG start codon and to achieve a T_m in the range of 58–62°C. At least 24 nucleotides of vector specific sequences are appended to 5′-end of both the forward [5′-TTAACTTTAAGAAGGAGTCTCTCCC] and reverse [5′-TTTTTCCATCCGCTAATGCTGCCCC] primers to facilitate Gibson assembly of the PCR product. Two SmaI restriction cleavage sites are located at the both ends of the ccdB gene. After digestion of the vector, no purification step is required for cloning process. Purified PCR product and the linearized vector will be assembled directly using Gibson assembly method described. The vector contains C-terminal CPD followed by hexa-histidine tag. The CPD-tag can be activated by adding inositol hexakisphosphate (InsP6). The protein of interest (POI) is then released

3. Thaw the reagents for PCR, reaction buffer, dNTPs, template DNA and PCR primers on ice.
4. Prepare PCR reaction mixture following Table 4 (*see* **Note 16**).
5. Put the PCR plate onto the thermal cycler and start.
6. Purify the PCR products using Qiagen Minelute PCR purification kit or a Millipore filter plate (*see* **Note 17**).
7. Verify the PCR products' size by running amplified product an agarose gel as detailed Subheading 3.1, **step 3** and determine concentration using a NanoDrop spectrophotometer (*see* **Note 18**).

3.3 Gibson Assembly Reaction

1. Gibson assembly reaction is a simple one-step cloning method [17–19]. Purified PCR product and SmaI-digested pCPD are mixed in the assembly master mix solution to perform cloning (*see* Fig. 2).
2. Thaw one aliquot of 15 μl assembly mixture on ice (*see* **Note 19**).
3. Prepare Gibson assembly reaction mixture in PCR tubes as described in Table 5 (*see* **Note 20**).

Table 3
Program of a thermal cycler for PCR

Cycle	Temperature	Time	Cycle repetitions
I	98 °C	1 min	1
II	98 °C	10 s	
	58 °C	30 s	30
	72 °C	30 s/kbp	
III	72 °C	10 min	1

Table 4
Preparation of master mix solution

PCR components	Single reaction	*N* reactions[a]	At the final reaction
Sterile deionized water	14.5 μl	14.5 × N μl	–
5 × buffer (HF)	5.0 μl	5.0 × N μl	1×
10 mM dNTPs	2.0 μl	2.0 × N μl	800 nM
Genomic DNA (50 ng/μl)	1.0 μl	1 × N μl	50 ng
Phusion DNA polymerase		0.5 × N μl	1 U
Master Mix volume	22.5 μl	23 × N μl	–
Forward primers (5 μM)	1 μl	1 μl	200 nM
Reverse primers (5 μM)	1 μl	1 μl	200 nM
Phusion DNA polymerase	0.5 μl		1 U
Total reaction volume	25 μl	25 μl/reaction	

[a]*N*, number of reactions. It is recommended to prepare 10% extra master mix solution for multiple PCRs. DNA polymerase can be added at the end of the PCR mix for single reactions or to the master mix for large numbers of reactions

4. Set up thermal cycler at 50 °C for 1 h and put the PCR tubes of assembly reaction mixture onto the thermal cycler and start (*see* **Note 21**). Cool to 4 °C until used.
5. Transformation to DH5α *E. coli* and plasmid preparation process are identical to previously reported protocols [20]. *E. coli* transformed with the fusion plasmid can be selected on agar plates containing 100 μg/ml ampicillin since the pCPD vectors carry the *bla* gene for ampicillin resistance. Sequence across SmaI site junctions and insert to confirm proper construction and lack of PCR generate errors.

Table 5
Gibson assembly reaction mixture

Components	Volume	Final concentration or quantity
Assembly master mixture	15 μl	1×
linearized vector backbone	× μl	78 fmol[a]
Purified PCR	× μl	78 fmol[a]
Water	Bring up the volume to 20 μl	

[a] *See* **Note 20** on establishing molar quantities

3.4 Bacterial Overexpression of CPD Fusion Proteins

Once the plasmid is assembled and verified by sequencing, the recombinant protein of interest fused to the CPD-tag is expressed in *E. coli*. Expression of CPD-fusion proteins can be done by any standard laboratory method for overexpression of genes under control of the T7 promoter [21, 22] (*see* **Note 7**). Only details specific for expression of CPD fusions are provided here.

1. A recommended strategy is to transform the completed, sequence-verified plasmid to *E. coli* BL21(DE3) or similar strain engineered for induced expression of T7 RNA polymerase. Select several single colonies and inoculate broth culture media containing 100 μg/ml ampicillin. Grow to OD_{600} of 0.4–0.6 and induce the culture by adding 0.1–1 mM IPTG followed by incubation at 16–37 °C for up to 16 h.
2. Small scale trials should be conducted to establish the IPTG concentration, temperature, and time to achieve optimal expression of soluble protein. A feature of the CPD-tag is that it has been shown to enhance the solubility of proteins that are insoluble without the tag [11]. The CPD-tag adds 23 kD to the size of the protein of interest (*see* **Note 22**).
3. Large scale bacterial production can then be conducted under similar conditions and bacterial cells collected at 4 °C by centrifugation. Remove all media and the freeze pellets at −20 °C for storage and to improve cell lysis during thawing.

3.5 IMAC Purification of CPD-Fusion Proteins

The first stage of protein preparation is to purify the full-length CPD-tagged fusion protein. This can be done by standard methods for preparation of a bacterial cell lysate, such as by sonication or French Press, followed by standard IMAC purification of C-terminally 6×His-tagged protein. CPD fusion proteins have been successfully purified by batch strategies, such as using Clontech TALON® metal affinity resin, and by FPLC using Ni-NTA columns such as the GE Healthcare HisTrap FF columns [11]. Only details specific for purification of CPD fusions are provided here.

1. Phosphate buffers must not be used at any stage of protein purification. Phosphate, even at modest concentrations, can mimic $InsP_6$ to organize the folding of the CPD and stimulate its autoprocessing [5], reducing yield of the full-length CPD fusion protein. Tris buffers have been successfully used for purification of CPD fusion proteins. Buffer should be slightly alkaline (pH 7.4–8.4) adjusted to be above the pI of the protein fusion accounting for the addition of the CPD-tag (pI = 6.1, *see* **Note 22**).
2. Inclusion of divalent cations to purification buffers should be avoided if possible. Divalent cations, especially Ca^{2+}, can bind to $InsP_6$ and prevent its binding to the CPD during the tag removal stage. If Ca^{2+} is strictly necessary for protein stability, add more $InsP_6$ and increase the CPD autoprocessing incubation time.
3. Chemical protease inhibitors that work by covalently modifying cysteine residues, such *N*-ethylmaleimide, must not be added to lysate or buffers as the modification will permanently inhibit the CPD and prevent tag removal at later stages of purification [5]. However, protease inhibitors such as Roche cOmplete protease inhibitor cocktail can be safely used. The *V. cholerae* MARTX toxin derived CPD that is expressed by the pCPD vector is insensitive to noncovalent protease inhibitors [10]. EDTA also will not inhibit the autoprocessing reaction since a metal ion is not required for *V. cholerae* CPD activity, and may even stimulate its activity by chelating divalent cations.
4. If during the expression trial the CPD fusion protein is insoluble, it is possible to purify CPD fusion proteins under denaturing conditions such as 6 M urea. The concentration of urea should be reduced by dialysis below 1 M prior to autoprocessing. $InsP_6$ has been shown in some cases to fold CPD out of low concentration urea enabling release of the protein of interest, although this may increase the likelihood of the protein of interest becoming cut at unintended site by in *cis* or in *trans* CPD processing (*see* **Note 23**).

3.6 CPD-Tag Removal Method #1: In Solution Autoprocessing

1. Exchange the buffer of the protein eluted from IMAC into Buffer A1 (no imidazole) by size exclusion chromatography, desalting column or dialysis. As noted above, this buffer should not be modified to include divalent cations.
2. Measure the volume of the protein solution using a pipet or estimate the volume using markers on the side of a 15 or 50-ml screw-cap conical centrifuge tube. Tube should be no more than 50% full to allow some movement of fluid during incubation.

3. Add 1:1000 dilution of 100 mM phytic acid solution ($InsP_6$) to adjust the protein solution to 100 μM $InsP_6$. Wrap the top of the conical tube with Parafilm to seal (*see* **Note 23**).
4. Place the tube on its side, and gently shake or rock for 1 hour at room temperature. Alternatively, if the cleaved target protein is not stable at room temperature, place the tube at 4 °C and for 2–4 h or overnight. Do not use rapid shaking or end-to-end rotor as excessive mixing can denature the protein.
5. Equilibrate an Ni-NTA HisTrap column (or metal charged resin) with Buffer A1.
6. Load $InsP_6$ cleavage reaction onto column or resin and collect the flow through. While collecting, wash the resin with additional 5 column volumes to ensure complete elution of the cleaved target protein (Sample 1). These fractions should contain most of the target protein.
7. Wash with 2 column volumes Buffer A1 containing 20 mM imidazole (Sample 2) and then with 2 column volumes Buffer A1 containing 50 mM imidazole (Sample 3). Collect these fractions as they may contain some of the target protein.
8. Elute the CPD-tag fragment and any uncleaved protein with Buffer A1 containing 500 mM imidazole (Sample 4).
9. Mix 10 μl of Samples 1–4 with 10 μl 2× Laemmli SDS-PAGE buffer, heat to 90 °C for 2 min and separate proteins by SDS-PAGE and visualize using Coomassie blue stain using standard protocols [15, 16]. The CPD-tag should run at 23.5 kD in Sample 4.
10. Concentrate final protein, adjust to 10% glycerol if needed to stabilize during freezing, and store at −80 °C in aliquots.

3.7 CPD-Tag Removal Method #2: On the Column Autoprocessing

1. Exchange the buffer of the protein eluted from IMAC into Buffer A1 (no imidazole) by size exclusion chromatography, desalting column or dialysis. As noted above, this buffer should not be modified to include divalent cations.
2. Equilibrate an Ni-NTA HisTrap FF column with Buffer A1 and load full-length CPD fusion protein onto column. The full-length protein will bind to the resin (Sample 1). The pH of the buffer may need to be adjusted according to the pI of the full-length protein with the CPD-tag attached (*see* Subheading 3.5, **step 1**).
3. Fill the column with Buffer A1 containing 100 μM phytic acid solution ($InsP_6$). Stop flow over column. Leave column overnight at 4 °C.
4. Using a flow rate of no more than 0.5 ml/min, elute protein of interest by continuing the flow of buffer A containing 100 μM phytic acid across the column, collecting the eluted in fractions.

The slow rate facilitates ongoing cleavage of any unprocessed protein during elution.

5. Wash column with Buffer A2 containing 20 mM imidazole (sample 2) and with Buffer A3 containing 50 mM imidazole (sample 3). Collect these fractions, they might still contain some amount of the target protein. Elute the CPD-6×His and any uncleaved protein with Buffer A4 containing 500 mM imidazole buffer (Sample 4).
6. Mix 10 μl of each eluted fraction and Samples 1–4 with 10 μl 2× Laemmli SDS-PAGE buffer, heat to 90 °C for 2 min and separate proteins by SDS-PAGE and visualize using Coomassie Blue stain following standard protocols [15, 16].
7. Concentrate final protein, adjust to 10% glycerol if needed to stabilize protein during freezing, and store at −80 °C in aliquots.

4 Notes

1. This method specifically details use of the pCPD-*ccdB* vector [11]. The progenitor vector pCPD described in the same paper is also compatible with all methods described here except that the SmaI-digested vector must be purified by gel extraction prior to use in Gibson assembly.
2. When designing primers against gene sequence for protein of interest, it is important to provide an ATG start codon at the start as the vector does not introduce a start codon. Also be sure to exclude the stop codon in the design for translation through to the CPD-tag.
3. PCR primers are designed such that gene specific sequences are followed by vector specific regions at the SmaI site. For pCPD, vector specific region sequences, 5′-TTAACTTTAAGAAGGA GTCTCTCCC and 5′-TTTTTCCATCCGCTAATGCTGC CCC are added to forward and reverse primers, respectively.
4. Prepare primer stocks at the concentration of 100 μM, and prepare dilution of working primer concentration of 5 μM by adding 5 μl of 100 μM stock to 95 μl deionized distilled water.
5. While the method described here is for PCR amplified DNA, the method is compatible with synthetic DNA molecules like gBlocks® (Integrated DNA Technologies) that should be designed with the same end extensions and parameters as for PCR primers in **Notes 2** and **3**.
6. Prepared 2× Gibson Assembly Master Mix also commercially available and convenient to use (New England Biolabs). The reaction size recommended can be reduced by half increasing

cost efficiency. However, for groups generating large number of clones or expecting frequent use, a self-mixed solution of enzymes and buffers as detailed in this protocol is more cost efficient.

7. Use of two different bacterial strains for cloning and overexpression is essential for success with this protocol. BL21(DE3) is a poor bacterial strain for recovery of ligated DNA and preparation of DNA for sequencing. Recovery of the Gibson assembly reaction should be done is strain compatible with DNA cloning such as DH5α. It is important that the cloning recovery bacterial cell not have a *gyrA* mutation as these cells can replicate vectors with *ccdB*.

   For overexpression of the protein fusion, BL21(DE3) produces the T7 RNA polymerase required for transcription of the fusion gene is encoded on a lambda prophage present on the *E. coli* chromosome. This T7 *gene1* is controlled by the *lacUV5* promoter that is normally repressed by the LacI repressor, which is also encoded on the prophage. Thus, expression of the T7 RNA polymerase and subsequently the fusion protein are inducible by the addition of IPTG.

8. Primers for sequencing the insert are standard primers: T7 promoter, 5′-TAATACGACTCACTATAGGG for forward sequencing and T7 terminator, 5′-CATCTGTCGTTGGTG TAACC for reverse sequencing.

9. This purification strategy is compatible with any method commonly used broth media for bacterial gene expression, including Luria–Bertani Broth, Terrific Broth, or supplemented minimal media. In particular, CPD fusions have been used to purify media for incorporation of selenomethionine or $^{15}N/^{13}C$ for structure determination. Base media should be sterilized by autoclaving in the culture flask. Flask should contain media volume of no more than quarter the volume of the flask for adequate aeration during growth with shaking. Cool media to room temperature and then add supplements including antibiotic.

10. For 2 M Tris–HCl Stock solution is made by dissolving 242.2 g Tris Base in 800 ml ddH_2O. Add 42 ml concentrate hydrochloric acid and fine-adjust to pH 8.0 using pH meter and bring up volume to 1-L using graduate cylinder. For 5 M NaCl stock solution, dissolve 292.2 g NaCl in 800 ml ddH_2O. Bring up volume to 1-L using graduate cylinder.

11. Prepare 2 M imidazole stock. Dissolve 136.15 g imidazole in 800 ml dH_2O. Adjust pH to 8.0 with HCl and bring final volume to 1 L. Keep stored in an amber bottle or wrap bottle in foil to protect from light.

12. SmaI digestion of pCPD-*ccdB* result in two pieces of DNA fragments: vector backbone and the *ccdB* gene. Although there might be partial digested vector or uncut vector in the reaction, these will still contain the *ccdB* gene. Thus, the fragments are lethal to *E. coli* after transformation and bacterial colonies will not form. Therefore, gel purification is not necessary when using the pCPD-*ccdB* vector. An addition advantage is the sizes of linearized vector backbone and partially digested pCPD-*ccdB* can be easily distinguishable on an agarose gel during verification of complete restriction digestion.
13. SmaI has a half-life of 15 min at 37 °C and thus increasing temperature above 25 °C can decrease cutting efficiency.
14. The size of the vector pCPD-*ccdB* is 7215 bp (2,227,816.9 g/mol) and pCPD is 5390 bp (1,665,311 g/mol). For example, 100 ng/μl of pCPD-ccdB is 44.9 fmol/μl (nM) and after digestion, the concentration of pCPD and *ccdB* are equally 44.9 nM. Since it is not necessary to purify the linearized vector backbone, the concentration of the uncut vector before restriction digestion reaction can be used to calculate vector backbone addition to the Gibson assembly reaction.
15. The extension (or elongation) time of the PCR is based on 30 s/kb as described in Table 3. When a batch of PCRs is set up, the largest target gene in the PCR batch should be used to set up the extension time. In general, it is recommended to make a PCR batch with a similar length of amplicons.
16. There are two options to prepare master mix solution for PCR. For small numbers of PCRs such as 1–8 reactions, master mix can be prepared with water, 5× buffer, dNTPs and genomic DNA (when the amplicons are in a same genomic DNA). Then, the master mix is split into PCR tubes. After adding PCR primers into the master mix, add 0.5 μl of Phusion DNA polymerase at the last step. For large numbers of PCRs such as 96 reactions, master mix is prepared with water, 5× buffer, dNTPs, genomic DNA and Phusion polymerase. Then the master mix is split into PCR tubes or a plate. Add 2 μl of combined-primer solution to each well using a multichannel pipette.
17. Using a MinElute column: 125 μl Buffer PB is added to the 25 μl PCR and apply on a MinElute column in a 2 ml collection tube. Centrifuge for 1 min at 13,000 × *g* of Eppendorf model 5415D for the sample to pass through the column. Wash the column with 750 μl Buffer PE by adding on the column and centrifugation for 1 min. Centrifuge the column for additional 1 min to dry the column completely. Replace the collection tube to a new microfuge tube and add 10 μl water and wait for 5 min. And then, centrifuge the column for 1 min to collect purified PCR.

For 96-wll PCR plate format: Transfer PCR products into a 96-well Millipore filter plate. Place plate on a vacuum system for 10 min or until the wells on the plate are dry. Resuspend PCR products in 40 μl of sterile deionized distilled water, shake for 20 min on a bench shaker at room temperature, and transfer to a new 96-well plate.

18. When handling small numbers of samples, it is easy to use agarose gels to analyze PCR products. Alternatively, an automated capillary electrophoresis system such as LabChip-GXII (PerkinElmer) can be used. The automated system is useful for large numbers of samples. The automated system determines concentration as well as size of DNA samples automatically. If the system is not available, use a spectrophotometer plate reader to determine PCR amplicon yields and concentration.
19. The assembly master mixture is 1.33× concentration. 15 μl of the master mixture is combined with 5 μl DNA mixture to perform cloning. The reaction mixture volume can be reduced to 10 μl (7.5 μl assembly master mixture and 2.5 μl DNA mixture).
20. Equal molar quantities of linearized vector backbone and purified PCR product are added in 15 μl assembly master mix and bring up the volume to 20 μl with deionized distilled water Table 5. describes the quantity of each DNA in 20 μl assembly reaction is 78 fmol, which is equivalent to 1 μl of 100 ng/μl 6 Kbp plasmid DNA. The quantity range of DNA can be used for the Gibson assembly reaction is 8–80 fmol in 20 μl assembly reaction. In 10 μl assembly reaction, recommended concentration of DNA fragments to be assembled is 25–50 fmol each. If the DNA concentration is too high to aliquot the desired quantity (if it is less than 0.5 μl), prepare dilution in deionized distilled water.
21. The reaction mixtures are required start reaction at 50 °C immediately. The assembled DNA can be transformed immediately as in Subheading 3.4 or stored at −20 °C for several days.
22. The CPD-tag contains the DNA sequence from *V. cholerae rtxA1* (*vc1451*) corresponding to codons for Leu3428–Gly3637 according to the correct 4545 amino acid annotation by Lin et al. [23, 24] followed by six histidines, not the incorrect 4558 amino acid annotation found on most databases. The protein sequence of the CPD-tag is: GAAL/ADGKILHN QNVNSWGPITVTPTTDGGETRFDGQIIVQMENDPVVA KAAANLAGKHAESSVVVQLDSDGNYRVVYGDPSKLDG KLRWQLVGHGRDHSETNNTRLSGYSADELAVKLAKFQ QSFNQAENINNKPDHISIVGCSLVSDDKQKGFGHQFIN

AMDANGLRVDVSVRSSELAVDEAGRKHTKDANGDWV QKAENNKVSLSWDAQGHHHHHH*

Processing occurs between the double-underlined L/A. The uncleaved CPD-tag is 219 amino acids with a theoretical pI of 6.09 and size of 23,816.2 Da. The cleaved CPD-tag is 215 amino acids with a theoretical pI of 6.09 and size of 23,503.8 Da. Cleavage leaves a sequence of GAAL at the C-terminus of the protein, which has less interference than the highly charged 6×His-tag in biochemical reactions.

23. The CPD is able to be reactivated for processing in trans and thus may for some proteins process available Leu-Xaa residues if found in unfolded regions of the protein of interest. Secondary site processing can be inhibited by addition of 100 μM L-leucine chloromethyl ketone hydrochloride (Sigma-Aldrich) although the initial cleavage reaction may also be impeded reducing overall efficiency of recover of the protein of interest [5].

Acknowledgments

This work funded in whole or in part with Federal funds from the NIAID, NIH, DHHS, under Contracts Nos. HHSN272201200026C and HHSN272201700060C. Additional funding from the Northwestern Medicine Catalyst Fund (to K.S.), NIAID Institutional NRSA Training Grant T32AI007476 (to P. W.) and PanCan/FNCLR KRas Fellowship (to M.B).

Conflict of interest: K.J.F.S. holds patents on the use of CPD for autoprocessing of fusion proteins.

References

1. Satchell KJ (2015) Multifunctional-autoprocessing repeats-in-toxin (MARTX) toxins of *Vibrios*. Microbiol Spectr 3:3. https://doi.org/10.1128/microbiolspec.VE-0002-2014
2. Gavin HE, Satchell KJ (2015) MARTX toxins as effector delivery platforms. Pathog Dis 73: ftv092
3. Lupardus PJ, Shen A, Bogyo M et al (2008) Small molecule-induced allosteric activation of the *Vibrio cholerae* RTX cysteine protease domain. Science 322:265–268
4. Prochazkova K, Satchell KJ (2008) Structure-function analysis of inositol hexakisphosphate-induced autoprocessing of the *Vibrio cholerae* multifunctional autoprocessing RTX toxin. J Biol Chem 283:23656–23664
5. Prochazkova K, Shuvalova LA, Minasov G et al (2009) Structural and molecular mechanism for autoprocessing of MARTX Toxin of *Vibrio cholerae* at multiple sites. J Biol Chem 284:26557–26568
6. Egerer M, Satchell KJ (2010) Inositol hexakisphosphate-induced autoprocessing of large bacterial protein toxins. PLoS Pathog 6: e1000942
7. Shen A, Lupardus PJ, Albrow VE et al (2009) Mechanistic and structural insights into the proteolytic activation of Vibrio cholerae MARTX toxin. Nat Chem Biol 5:469–478
8. Shen A, Lupardus PJ, Morell M et al (2009) Simplified, enhanced protein purification using an inducible, autoprocessing enzyme tag. PLoS One 4:e8119

9. Satchell KJ (2010) Cysteine protease autoprocessing of fusion proteins US Patent No. 8,257,946 B2. Date of issue: Sep. 4, 2012
10. Sheahan KL, Cordero CL, Satchell KJ (2007) Autoprocessing of the *Vibrio cholerae* RTX toxin by the cysteine protease domain. EMBO J 26:2552–2561
11. Biancucci M, Dolores JS, Wong J et al (2017) New ligation independent cloning vectors for expression of recombinant proteins with a self-cleaving CPD/6xHis-tag. BMC Biotechnol 17:1
12. Eschenfeldt WH, Makowska-Grzyska M, Stols L et al (2013) New LIC vectors for production of proteins from genes containing rare codons. J Struct Funct Genomics 14:135–144
13. Lee PY, Costumbrado J, Hsu CY et al (2012) Agarose gel electrophoresis for the separation of DNA fragments. J Vis Exp 62:3923. https://doi.org/10.3791/3923
14. Makovets S (2013) Basic DNA electrophoresis in molecular cloning: a comprehensive guide for beginners. Methods Mol Biol 1054:11–43
15. Gibbins JM (2004) Techniques for analysis of proteins by SDS-polyacrylamide gel electrophoresis and Western blotting. Methods Mol Biol 273:139–152
16. Chakavarti B, Chakavarti D (2008) Electrophoretic separation of proteins. J Vis Exp 16:758. https://doi.org/10.3791/758
17. Gibson DG, Benders GA, Andrews-Pfannkoch C et al (2008) Complete chemical synthesis, assembly, and cloning of a *Mycoplasma genitalium* genome. Science 319:1215–1220
18. Gibson DG, Benders GA, Axelrod KC et al (2008) One-step assembly in yeast of 25 overlapping DNA fragments to form a complete synthetic *Mycoplasma genitalium* genome. Proc Natl Acad Sci U S A 105:20404–20409
19. Gibson DG, Young L, Chuang RY et al (2009) Enzymatic assembly of DNA molecules up to several hundred kilobases. Nat Methods 6:343–345
20. Kwon K, Pieper R, Shallom S et al (2007) A correlation analysis of protein characteristics associated with genome-wide high throughput expression and solubility of *Streptococcus pneumoniae* proteins. Protein Expr Purif 55:368–378
21. Samuelson JC (2011) Recent developments in difficult protein expression: a guide to *E. coli* strains, promoters, and relevant host mutations. Methods Mol Biol 705:195–209
22. Baneyx F (1999) Recombinant protein expression in *Escherichia coli*. Curr Opin Biotechnol 10:411–421
23. Lin W, Fullner KJ, Clayton R et al (1999) Identification of a *Vibrio cholerae* RTX toxin gene cluster that is tightly linked to the cholera toxin prophage. Proc Natl Acad Sci U S A 96:1071–1076
24. Dolores J, Satchell KJ (2013) Analysis of *Vibrio cholerae* genome sequences reveals unique *rtxA* variants in environmental strains and an *rtxA*-null mutation in recent altered El Tor isolates. mBio 4:e00624

INDEX

Gregory J. Miller (ed.), *Inositol Phosphates: Methods and Protocols*, Methods in Molecular Biology, vol. 2091, https://doi.org/10.1007/978-1-0716-0167-9, © Springer Science+Business Media, LLC, part of Springer Nature 2020

K

L

R

S

T

U

V

W

X

FSC
www.fsc.org
MIX
Papier aus verantwortungsvollen Quellen
Paper from responsible sources
FSC® C105338

If you have any concerns about our products,
you can contact us on
ProductSafety@springernature.com

In case Publisher is established outside the EU,
the EU authorized representative is:
Springer Nature Customer Service Center GmbH
Europaplatz 3, 69115 Heidelberg, Germany

Printed by Libri Plureos GmbH
in Hamburg, Germany